Meaning from Data: Statistics Made Clear

Part II

Professor Michael Starbird

THE TEACHING COMPANY ®

PUBLISHED BY:

THE TEACHING COMPANY
4151 Lafayette Center Drive, Suite 100
Chantilly, Virginia 20151-1232
1-800-TEACH-12
Fax—703-378-3819
www.teach12.com

Printed in the United States of America

ISBN 1-59803-147-3

Michael Starbird, Ph.D.

University Distinguished Teaching Professor of Mathematics,
The University of Texas at Austin

Michael Starbird is a professor of mathematics and a University Distinguished Teaching Professor at The University of Texas at Austin. He received his B.A. degree from Pomona College in 1970 and his Ph.D. in mathematics from the University of Wisconsin, Madison, in 1974. That same year, he joined the faculty of the Department of Mathematics of The University of Texas at Austin, where he has stayed except for leaves as a Visiting Member of the Institute for Advanced Study in Princeton, New Jersey; a Visiting Associate Professor at the University of California, San Diego; and a member of the technical staff at the Jet Propulsion Laboratory in Pasadena, California.

Professor Starbird served as Associate Dean in the College of Natural Sciences at The University of Texas at Austin from 1989 to 1997. He is a member of the Academy of Distinguished Teachers at UT and chairs its steering committee. He has won many teaching awards, including a Minnie Stevens Piper Professorship, which is awarded each year to 10 professors from any subject at any college or university in the state of Texas; the inaugural award of the Dad's Association Centennial Teaching Fellowship; the Excellence Award from the Eyes of Texas, twice; the President's Associates Teaching Excellence Award; the Jean Holloway Award for Teaching Excellence, which is the oldest teaching award at UT and is presented to one professor each year; the Chad Oliver Plan II Teaching Award, which is student-selected and awarded each year to one professor in the Plan II liberal arts honors program; and the Friar Society Centennial Teaching Fellowship, which is awarded to one professor at UT annually and includes the largest monetary teaching prize given at UT. Also, in 1989, Professor Starbird was the Recreational Sports Super Racquets Champion.

The professor's mathematical research is in the field of topology. He recently served as a member-at-large of the Council of the American Mathematical Society and currently serves on the national education committees of both the American Mathematical Society and the Mathematical Association of America.

Professor Starbird is interested in bringing authentic understanding of significant ideas in mathematics to people who are not necessarily mathematically oriented. He has developed and taught an acclaimed class that presents higher-level mathematics to liberal arts students. He wrote, with co-author Edward B. Burger, *The Heart of Mathematics: An invitation to effective thinking*, which won a 2001 Robert W. Hamilton Book Award. Professors Burger and Starbird have written a book that brings intriguing mathematical ideas to the public, entitled *Coincidence, Chaos, and All That Math Jazz: Making Light of Weighty Ideas*, published by W.W. Norton, 2005. Professor Starbird has produced two previous courses for The Teaching Company, one entitled *Change and Motion: Calculus Made Clear* and a second one with collaborator Edward Burger entitled *The Joy of Thinking: The Beauty and Power of Classical Mathematical Ideas*. Professor Starbird loves to see real people find the intrigue and fascination that mathematics can bring.

Acknowledgments

These lectures were prepared in collaboration with Thomas Starbird, Ph.D., a principal member of the technical staff at the Jet Propulsion Laboratory, Pasadena, California, and Jennifer Kaplan, a Ph.D. student in statistics education at The University of Texas at Austin. We three worked together on the concept, design, and details of the entire course. I would also like to thank the following people for discussions and ideas: Joe Gallian, Terry Kahn, Mary Parker, Tony Petrosino, James Scott, and Ann Watkins. Thanks to Lucinda Robb, Noreen Nelson, Pamela Greer, Alisha Reay, and others from The Teaching Company, not only for providing excellent professional work during the production of this series of lectures but also for creating a supportive and enjoyable atmosphere in which to work. Finally, thanks to my family, Roberta, Talley, and Bryn, for their special encouragement.

Table of Contents
Meaning from Data: Statistics Made Clear
Part II

Meaning from Data: Statistics Made Clear

Scope:

A statistical fact: On average, each American has one testicle and one ovary.

Should we take cholesterol-lowering medication? Evidence for and against is presented to us in the form of data and statistical conclusions. Should we buy stocks or sell? Much of the information we use to make the decision is based on numerical data. Will it rain tomorrow? Will the real estate market rise or fall? How good a student will Mr. Jones be, if admitted? Should we buy lottery tickets when the jackpot gets really big? Should a coach leave a player in the game when he's in a slump? How can we tell if gender discrimination influenced college admissions procedures? Trying to understand the economy, the weather, school systems, grading, the quality of products, risk, measurements of everything, social trends, marketing, science, and most practical aspects of our world fundamentally involves coming to grips with data.

The trouble with data is that data do not arrive with meaning. Data are value-free and useless or actually misleading until we learn to interpret their meaning appropriately. Statistics provides the conceptual and procedural tools for drawing meaning from data.

Analyzing data correctly is one of the most powerful tools that we have for understanding our world. But it is a two-edged sword. Mark Twain attributed to Benjamin Disraeli perhaps the most famous quip about statistics: "There are three kinds of lies: lies, damned lies, and statistics." But an apt rejoinder is: "It is easy to lie with statistics, but it is easier to lie without them." In this course, we will see the two sides of data—their uses and their misuses.

We will learn basic principles and ideas of statistics and understand how they can bring meaning to data. We will learn about probability and the central role it plays in understanding the meaning of statistics. One of the great ideas of modern quantitative analysis of our world is that the uncertain and the unknown can be described quantitatively. Random events show global trends in the aggregate, and probability and statistics can help us describe and measure those trends.

We present statistics by isolating two major challenges: (1) How can we describe and draw meaning from a collection of data when we know all the pertinent data? (2) How can we infer information about the whole population when we know data about only some of the population (a sample)? These two questions form the structural backbone of our approach.

The challenge of describing a collection of data when we know all the data arises, for example, when we have complete records of all students who have ever attended a given university. We know the incoming Scholastic Aptitude Test (SAT) scores and high school class rank of all students, and we know their grade point averages (GPAs) on graduation. We can ask and answer many questions regarding those data. Perhaps we would like to know some summary information, for example, the mean GPA or the range of SAT scores. Maybe we would like to describe how well the SAT scores and high school rank-in-class predict the students' future performance. Describing income data, age data, sports statistics, and a myriad of other examples all present us with the challenge of taking a mound of figures and assembling them in a fashion from which we can glean meaning.

The second challenge is the challenge of statistical inference. Suppose we take a poll of 1,000 voters before an election to find out how they will vote. We really want to know how the 100 million voters will vote in the next election. How confident can we be that the opinions of the 1,000 voters we ask really do reflect the opinions of the 100 million voters who will vote in the election? That is one of the challenges of statistical inference. Predicting the future weather given information about past weather, deducing whether a new drug is efficacious, guessing the future performance of the stock market, and doing scientific experiments on a few mice and drawing conclusions about all animals are all examples of the statistical challenge of inferring conclusions about the whole population when we have information about only a sample of the population.

Part I of this course, Lectures One through Twelve, introduces the concepts of statistics. Typically, several different application areas are used to illustrate each statistical concept. Part II, Lectures Thirteen through Twenty-Four, is organized by application area. Typically, several different statistical concepts are introduced and used in each application area. Both parts of the course are full of interesting and entertaining examples from all corners of our world—business and economics, medicine, education, sports, social science, and many more areas. For example, we will see how statistics was used to estimate the number of German tanks in World War II from the serial numbers of captured tanks, and we will see how a statistical analysis provides strong evidence concerning the disputed authorship of 12 of *The Federalist Papers*. Statistics offers unrivaled scope for connecting inherently intriguing mathematical ideas to the real world.

A typical statistics course in college emphasizes various technical tests. Students emerge with the impression that statistics amounts to plugging data into a formula. Although we will introduce important statistical formulas, this course emphasizes the logical foundations and underlying strategies of statistical reasoning. We describe why randomness lies at the heart of statistical reasoning. We explain what it means when the headlines blare, "Candidate A to get 59% of the vote with a + or – 3% margin of error." We differentiate between *statistically significant* and *significant*.

Our goal is to convey an authentic understanding of one of the most useful, powerful, and pervasive modes of reasoning employed in the world today. We will see why statistics will become increasingly important as technological advances continue to bring larger data sets and more detailed techniques of analysis within the range of practicality.

Note: Although the data used in this booklet are often real, some have been created to illustrate particular statistical concepts.

Lecture Thirteen
Law—You're the Jury

Scope:

The law provides many examples of instances in which statistical data and inferences are central to the decisions being made. We'll look at two main examples of courtroom drama where we imagine ourselves on the jury. We'll present the facts, and it will be up to us to determine what conclusions are appropriate to draw. We begin with a hit-and-run accident. The star witness gives crucial testimony with a high degree of credible confidence. But our interpretation of the significance of the testimony requires a little more thought. By viewing the witness's observations as a sample, we can understand the significance of the evidence in a quantitative way. The second courtroom example involves a gender-discrimination case. Do the data support an allegation of illegal discrimination on the basis of gender or do the data suggest a mere random association with gender? This example illustrates a surprising statistical anomaly known as *Simpson's Paradox*.

Outline

I. This lecture begins the second part of the course, in which we look at applications of statistics to different situations and subject areas. This lecture deals with the law.

- **A.** Statistical data and inferences are used frequently in the law and in making legal decisions.
- **B.** We will present the evidence; you determine the verdict.

II. Case 1 is a hit-and-run accident.

- **A.** Mr. Jones witnessed a hit-and-run accident involving a cab.
- **B.** Mr. Jones stated that he thought the cab was blue.
- **C.** The jury has come to the conclusion that a guilty verdict hangs on whether or not Mr. Jones's testimony implies that the probability that the cab was blue is over 50%.

1. We did some experiments to determine how accurate Mr. Jones's vision was. He was able to correctly identify the blue cabs as blue 80% of the time and the green cabs as green 80% of the time.
2. The prosecutor summarized the situation by saying that Mr. Jones is 80% sure the cab was blue; therefore, the jury should convict.

D. Some additional information was presented:

1. There are exactly 100 cabs in the city.
2. 90 are green cabs and 10 are blue.
3. Let's do the following thought experiment: Suppose instead of having one accident, all 100 cabs in the city simulated this accident. Each one did it once. How many times would Mr. Jones have testified that the cab was blue?
4. Following the logic, we see that he would testify "blue" 26 times, but only 8 of those times would the cab actually be blue.

III. Another case that demonstrates the same statistical issue about testing for rare events arises when considering random drug testing in a company.

A. Suppose the company has 280,000 employees, of whom 500 employees actually use the illegal drugs that are the target of this policy.

B. Suppose the drug test will correctly read positive for 95% of people who actually use those drugs; thus, 475 employees who use drugs would receive a positive test result.

C. Now suppose that the test gives a false positive 1% of the time; of the 279,500 employees who do not use drugs, 1% will get a false-positive result. That is, the drug test will read positive for 2,795 employees who do not, in fact, use drugs.

D. The total number of employees whose drug tests are positive will be 475 + 2,795 = 3,270 employees.

1. Therefore, if an employee gets a positive test result, his or her chance of actually using drugs is $\frac{475}{3270}$, or less than 15%.

2. Thus, there is a great danger of inappropriate firings or accusations based on positive drug testing.

IV. We now turn to gender discrimination.

A. To determine discrimination by gender, race, religion, or age, it is natural to look at data about treatment of specific groups.

B. Let's consider a case of admissions to a program at a university. For the purposes of this example, let's assume that 1,000 men applied and 1,000 women applied to the program and that all 2,000 applicants had exactly the same qualifications. Here are the facts:

C. There was a 70% acceptance rate for the men and only a 40% acceptance rate for the women.

1. It appears that the women were clearly discriminated against, and of course, we are outraged.

2. The chi-square test can identify the acceptance rate you would expect from random chance alone.

D. But let's look at the case more deeply. Suppose the total program to which the 2,000 applied actually had two subprograms.

1. One subprogram was an Excellent Program to which 200 men applied and 800 women applied.

2. The other subprogram was a Mediocre Program to which 800 men applied and 200 women applied.

E. Of the 200 men who applied to the Excellent Program, 20% were accepted.

F. Of the 800 women who applied to the Excellent Program, 25% were accepted.

G. In the Mediocre Program, 800 men applied and 82.5% were accepted, while 200 women applied to the Mediocre Program and 100% were accepted.

H. In each subprogram, a higher percentage of women were accepted. This situation is an anomaly, because overall, it appeared that the women were being discriminated against; however, looking at the subprograms we see a different picture.

I. This scenario is an illustration of a phenomenon known as *Simpson's Paradox*, that is, a situation where both subprograms indicate that the women are being treated better, yet overall, the men appear to be treated better.

V. Given that the women actually had a higher acceptance rate in each subprogram, let's think about how we would decide whether the differences in acceptance rates are serious enough to be viewed as clear discrimination or whether the differences could be reasonably accounted for as the result of simple random luck.

A. Let's look at a table that records the data:

Excellent Program	Accept	Reject	Total	Accept Rate
Men	40	160	200	20%
Women	200	600	800	25%
Total	240	760	1,000	24%

Mediocre Program	Accept	Reject	Total	Accept Rate
Men	660	140	800	82.5%
Women	200	0	200	100%
Total	860	140	1,000	86%

B. How rare an event would it be to have one acceptance rate as much as 5% less than the other?

C. To determine how surprised we should be at getting a 5% difference in acceptance rates, we can look at a normal probability curve that tells us the probability of different deviations.

VI. The gender discrimination case illustrated a statistical anomaly known as *Simpson's Paradox*.

A. Simpson's Paradox is an example of a possible effect of a lurking variable.

B. In that case, the lurking variable was the existence of the subprograms.

VII. Two other examples of legal issues arose during the O. J. Simpson trial.

A. After evidence had been presented that O. J. had been guilty of wife beating, his lawyer, Johnnie Cochran, presented evidence that only 1 in 1,000 wife beaters went on to kill their wives.

B. Given that O.J. beat his wife, Cochran argued, there is only a 1 in 1,000 chance that he went on to commit the murder.

C. There are two fallacies here. The first regards the relative frequency of murders by wife-beaters compared to murders by non-wife-beaters.

1. The fact that 1 in 1,000 wife-beaters go on to murder their wives needs to be compared with the rate of *non*-wife-beaters who murder their wives.

2. The rate of murder among non-wife-beaters is much smaller than among wife-beaters; thus, evidence of wife-beating increases the likelihood of murder relative to others in the population.

D. The second fallacy in Cochran's argument is that a wife was actually murdered. The relevant question would be, if a wife is murdered, what is the probability that she had previously been beaten?

E. The idea of using wife beating as exculpatory evidence is ridiculous.

F. Statistics never proves ridiculous conclusions.

VIII. Another statistical issue about trials is the following: DNA evidence is not in itself damning if the DNA was used to find the culprit.

A. Suppose everyone's DNA were on file. A crime is committed, and the perpetrator's DNA is found on the scene. One out of a million has a specific matching DNA type. The data bank is combed, and someone matching the DNA type is arrested. At the trial, the prosecutor says, "There is only a one in a million chance that the DNA type would match." But that is bad reasoning, because DNA type was used to arrest the person.

B. Instead, if there are 10 in the large city with that DNA type, then the probability of guilt would be only 1 in 10.

IX. The law provides many opportunities in which statistics plays a significant role in the dispensation of justice.

A. In this lecture, we saw how a witness's testimony was not as compelling as it at first appeared.

B. We saw that universal or random drug tests present problems because of false positives.

C. In the gender-discrimination case, we saw an example of Simpson's Paradox; we also saw how the principles of hypothesis testing helped us to interpret evidence as indicative of discriminatory practices or not.

D. In the next two lectures, we turn to statistical anomalies involved with voting.

Readings:

Donald A. Berry, *Statistics: A Bayesian Perspective*.

David S. Moore and George P. McCabe, *Introduction to the Practice of Statistics*, 5th ed.

Questions to Consider:

1. Suppose two eyewitnesses had identified the cab as blue in the car accident case. What would be the probability that the cab was actually blue? Would you convict in that case?
2. Is it possible to devise an example in which a higher percentage of men than women are admitted to a program, but upon looking at two subprograms, a higher percentage of women are accepted than men in both subprograms, and upon looking at two *sub*-subprograms in each of the subprograms, in all four sub-subprograms, the men have a higher acceptance rate than the women?

Lecture Thirteen—Transcript
Law—You're the Jury

Welcome back to *Meaning from Data: Statistics Made Clear*. With this lecture, we begin the second part of the course, which, as advertised, looks at applications of statistics to different areas. In this lecture, we are going to look at applications to the law. Before I begin, let me just say that it's not the intent of these lectures to give any sense of overview of the potential applications of statistics to the different areas. Instead, what we're doing is just picking out some interesting individual applications and looking at them, with no attempt to give a global view.

But, certainly, in the law, there are many places where statistical data and inferences are used to help in making legal decisions. In this lecture, we're going to look at two cases where you will be the jury. In fact, there's a good reason to think of the law as the first application area to look at in this second part of the course about statistics; because really, ultimately, statistics is a reasoning tool. It's a tool by which we can make decisions about the world. It helps us to evaluate evidence and construct persuasive arguments about things. So, the idea of imagining ourselves as on the jury, I think really puts us in the right frame of mind to think about the real purpose of statistical analysis.

In this case, we're going to present some evidence, and interpret the evidence to see how significant that evidence may be to make a decision. In particular, in the examples that we'll give today, you'll see that the evidence may not actually have the implications that first appear.

We'll begin with Case Number One. Remember, you are the juror. Here is the scenario. This is a hit-and-run accident case. The scenario is that at twilight one night, Mr. Jones was taking his garbage out to the side of the street, and he looked down the street, and a couple blocks away, he saw an accident. A taxicab drove into an intersection, hit another car, and then drove off.

A cabbie has been accused of this crime and is the defendant. The evidence is quite convincing up to this point, but the jury, in weighing the balance of guilt, has decided that the guilty verdict will hang on the testimony of Mr. Jones, this eyewitness. In particular, it's going to hang on the issue of whether Mr. Jones saw a green cab or a blue cab. Here's the way the evidence was presented and then you're going to evaluate the import of the evidence.

The prosecutor asks Mr. Jones, "What did you see?" Mr. Jones said, "I thought that the cab was blue. I definitely saw a cab," everybody agrees it was a cab, "and I think it was blue. I'm pretty sure it was blue." You know how lawyers are; they'll jump on that. "What do you mean pretty sure? How sure are you?"

Mr. Jones, fortunately, having taken some statistics courses, was ready with a response to this because he said, "I'll tell you how sure. I'm 80% sure because here's what I did. We took thousands of trials where we randomly chose blue and green cabs. At twilight, I stood out by my garbage can and I looked at exactly the same place, under exactly the same conditions, the exact same time at night, and we did this experiment many times over. It turned out that, in 80% of the times, I was able to say correctly that it was blue when it was blue; and 80% of the time when it was green, I was able to identify that it was green." The prosecutor summarized the situation by saying that Mr. Jones is 80% sure that the cab was blue and, therefore, the jury should convict.

Well, that sounds pretty persuasive. But the defense lawyer hasn't started yet. The defense lawyer then comes up and has some additional information and poses a hypothetical question to the jury. The additional information is that it turns out that there are exactly 100 cabs in this particular city—90 of the cabs are green, and 10 of the cabs are blue. Now, everybody agrees that Mr. Jones is 80% accurate in correctly identifying green cabs as green or blue cabs as blue, in the conditions of the moment when he was the witness.

But let's do the following thought experiment: Knowing that there are 90 green cabs in the city, and 10 blue cabs in the city, before we begin thinking about the guilt of this cabbie, we have to be open-minded and imagine that it's equally likely that any one of those 100 cabs might have been the guilty cab. Well, the eyewitness is correct 80% of the time. Let's just see the implications of that.

If we did the thought experiment of all 100 cabs in the city having this accident, and asking ourselves what would the witness, Mr. Jones, testify over those entire 100 different trials, here's what he would do. For the 90 times that the cab was green, since he's 80% correct, on 18 occasions, he would actually testify that he thought it was blue because, you see, he's wrong 20% of the time, and 20% of 90 is 18. On the other hand, if, in the 10 instances in which the cab actually was blue, he would say that the cab was blue 8 times, and 2 times, he'd say it was green.

So, in his testimony about the entire city of cabs—the entire 100 cabs—he would actually testify that it was blue in 26 of those testimonies. In other words, 18 times when the cab was green, he would say it was blue, and 8 times when the cab actually was blue, he would say it was blue. So, the total number of times that he would testify 'blue' is 26 times; but it actually was blue only 8 times. The probability that the cab actually is blue, given the testimony that he says it's blue, is actually only 8 out of 26, or 31%.

So, given this interpretation and this analysis, you on the jury now have a very different view of the likelihood of the cabs actually being blue and, in fact, you should acquit. It's rather interesting. So, this is an example where your initial impression about the cabs was changed. Sixty-nine percent of the cabs, when the witness would say they were blue, actually turned out to be green.

Let's look at another example that involves a similar statistical issue. This time, let's imagine that we're in a very large company—it has 280,000 employees, and we're concerned with drug use, particularly of employees, particularly for a very serious kind of a drug. And so, in the wisdom of the company, the company decides, "Well, why don't we think about instituting universal drug testing? Just test every single person in the company, and then we'll find out, detect, who is using this very dangerous drug."

Well, we need to look into the accuracy of the test—so what we do is we find a test, and we can actually measure the quality of the test by doing experiments with the test and seeing how often the test actually correctly identifies a drug user. Let's suppose that in this particular test that it is 95% accurate if the person uses drugs—that is, it says the person uses drugs 95% of the time that the person *does* use that drug—and 1% of the time, it will say that the person uses the drug, even though that person *does not* use the drug.

So, let's do some analysis and see what would happen if we instituted universal drug testing. All right? So, here we go; let's start looking at some data. Let's suppose that the reality is that about 500 people use this serious drug out of the 280,000 employees. So, the reality is that there are 279,500 non-users, and 500 users of this drug. Well, the drug test, as we said, is 95% correct for the drug users—it identifies them as users—and 99% correct for the non-users. Let's see what happens when we apply the test.

Applying the test to these non-users, most of the time—276,705 times—it correctly says they're non-users; but 1% of the time, it gives a false positive—2,795 times, it says the person is a drug user, even though that person is not. Among the users, 475 times, the test says that the user is a user when, indeed, that person is a user; and 25 times, it does not identify that person as a user.

Let's see what happens, overall, if we gave this test to everybody in this company, these entire 280,000 employees. Well, how many times would it come back positive, saying the person is a user? It would be these 2,795 times, plus the 475 times here for a total of 3,270 times, it came back as a positive test. However, among that group, only 475 people were actually users of this drug. That means that if an employee got a positive drug test, the actual correct interpretation of that positive result is that there's only a 14.5% chance that that person actually is a user.

So, there's a great danger of inappropriate accusations, or firings, if you instituted this policy. In fact, I think there would be a class-action suit against the use of this policy because you were condemning so many people for drug use when, in fact, only 15% of the people with a positive test actually used the drug. Specifically, 85% of the people who test positive for that drug are not actually users of that drug.

Next, let's turn to a different kind of a legal issue. This is an issue having to do with discrimination on the basis of gender. This is one of the types of cases that come up all the time. You want to know whether or not, in a hiring practice or acceptance to a university, if there are discriminatory practices that make it unfair in dealing with different categories of people. So, let's just look at an example and see whether or not we are going to be convinced by the data that this university has committed a serious crime of discrimination.

So, let's look at this particular example. Here is a hypothetical university admissions case where there are a total of 2,000 applicants to this particular university; 1,000 men apply, and 1,000 women apply. Let's assume, for the sake of argument, that all of these people are completely equally qualified for admission to the university. Of course, this is not realistic, but these are some of the kinds of things that you have to presume if you're going to make a determination about discrimination; you have to eliminate some of the variables. So, we'll eliminate all of the variables except for the gender.

So, we have 1,000 men and 1,000 women who apply to the school, and look what happened—70% of the men were accepted, and only 40% of the women were accepted. Well, we should be outraged. Or should we be outraged? How do we decide if we should be outraged at this blatant discrimination against women?

Well, we have to think. We have to do some analysis and say, "How rare an occurrence would it be for only 40% of women to be accepted and 70% of the men to be accepted if the overall acceptance rate, as it is here, is 55%?" In other words, let's say you had 2,000 poker chips, blue and red. Just by random luck alone, if you randomly selected 55% of the 2,000 chips, what's the chance that the difference in the proportion of blue ones that you had differed as much from the proportion of reds as we have here, 70% versus 40%?

Well, you see, that this is putting us in the realm of the kind of statistical and probabilistic analysis that we have become accustomed to in the last several lectures. Namely, we can look at all possible gatherings of 1,100 out of this 2,000 and see what fraction of those have a disproportionate balance that is recorded in this example.

In fact, there's a specific test that accomplishes this. Namely, what we do is we look at every box in our table, and we know that there's an expected value—namely 55% of the men. If 55% is the overall acceptance rate, then the expected value in this square would be 55% of the 1,000 men. Therefore, 550 would be the expected value here, and the rejection square would be 450. Similarly here, the numbers would be 550 and 450. So, those are the expected values. We can take the observed values minus the expected values, square it, and divide by the expected count, and add them up over the 4 squares, and this gives a statistic called the *chi-square statistic*.

All it's doing, as I previously said, is looking at what you would expect by random chance alone. Then you look up on a table to see how far, the number that you get, if it's a larger number—because of the fact that the expected value versus the actual values differ by a lot, then when you square them and add them up, you get bigger numbers—if you get a bigger number, then that's an indication that there is some sort of unlikeliness to the outcome.

In this case, the value of adding up those numbers is 182, and we see that the chance of that actually happening is extremely small, 2×10^{-16}. That means you put a decimal down, you put 15 zeros and a 2. That's the probably that there'd be that big of a disparity in our count.

So, it sounds like if we're on the jury, we're going to say there's great discrimination against those women, right? Okay, wait a minute though. I forgot to tell you something. What I forgot to tell you is that actually, the university had two subprograms. Although it's true that 2,000 people applied to the university, actually, there were two subprograms.

One of the subprograms is the Excellent Program, and it has higher standard than the other program, which is the Mediocre Program. Now, in this Excellent Program, 200 men applied and 800 women applied. The overall acceptance rate to this Excellent Program was only 24%; it's a more discriminating program than the overall university. Look what happened. The men had an acceptance rate of 20%, and the women had an acceptance rate of 25%. So, for the Excellent Program, the rate of acceptance for the women was actually higher than the rate of acceptance for the men.

Oh, wait a minute. Now let's look at the Mediocre Program. In the Mediocre Program, 800 men applied and 200 women applied. Look what happened there. For the men, 82.5% were accepted; but every woman who applied to this Mediocre Program was admitted.

Now, let's make sure you understand here what's happening. What this says is that overall—let's go back to the chart—overall, the acceptance rate of the men was 70%, and the acceptance rate of the women was 40%. But in each of the subprograms—in the Excellent Program, the acceptance rate of the women was higher than the acceptance rate of the men; and, also, in the Mediocre Program, the acceptance rate of the women was also higher than the acceptance rate of the men.

This possibility occurs because of the fact that there are two subprograms, and in the two subprograms, there's a different rate of acceptance. More women applied to the tougher subprogram and, consequently, the effect of this was this paradoxical feature that in both of the subprograms, it looks, in fact, as though the discrimination is going the other way, right? Because in the subprograms, here the women have a higher acceptance rate in both the Mediocre Program and in the Excellent Program. In both cases, the women have a higher acceptance rate.

This is an example of a phenomenon called *Simpson's Paradox*, which is where the overall percentages seem to go one way, and yet both of the subcategories go the other way. I think it's a fascinating issue.

Now, if you're on the jury, now you have to decide if you should be outraged that the men are being discriminated against in each of the subprograms. You see? How do we do that? Once again, we would use our statistical analysis to see how unlikely is it, if the overall acceptance rate is 24% to a program, and this many men apply—200—and this many women apply, what is the chance that the acceptance rates would be in this category? We can run the numbers and see that there would be a .14 chance. It would happen 14 times out of 100 just by luck alone that you would have that much of a disparity in the acceptance rates of the men and the women.

The same thing applies to the Mediocre Program. We can do the same kind of analysis and, in that case, we see that it's extremely unlikely that we would have as much disparity in the acceptance of the men and the women. So, in this subprogram, where all of the women were accepted, it would be very unlikely, by luck alone, that when you chose 86% of the total applicants, that the imbalance of the proportion of the women being accepted to the men being accepted would be as great as it is. So, this would be evidence, and if you wanted to convict, you would want to say that, in fact, there was discrimination against the men.

So, this is a wonderful example where your first look at that data appeared to say one thing; but the second look at the data appeared to say something entirely different. That's Simpson's Paradox.

I'd like to discuss two other jury issues that came up during the O.J. Simpson trial. As you'll recall, when O.J. Simpson was on trial, Johnnie Cochran was the lead defense lawyer. During the course of the trial, there was some interesting statistical evidence educed, usually by the prosecution, but sometimes by the defense as well. This is an example of a statistic that was presented by the defense.

At the trial, evidence was presented that O.J. Simpson, on previous occasions, had beat his wife—you may remember this, that he beat his wife—and that was part of the evidence; there were 911 calls and so on that showed that he had beat his wife. The prosecution had put forward this evidence as an indication of his habits of violence.

The defense made the following argument. They said, "Well, we've looked at statistics, and we've found that only 1 out of 1,000 people who beat their wives go on to murder them." So, they presented this argument as exculpatory evidence that, in fact, O.J. Simpson would probably not be guilty because, look, only 1 out of 1,000 people who beat their wives actually go on to murder them.

Now, I bring this up for a couple of reasons. The first one is, if you hear something that sounds ridiculous and is claimed to be the result of a statistical argument, then you should be very skeptical of it. It probably is ridiculous. Of course, it's ridiculous. The argument that he beat his wife is exculpatory to the idea that he murdered his wife, that is, on its face, ridiculous. Let's look at why the statistics don't actually mean what it was presented to the jury as meaning.

First of all, if 1 out of 1,000 people who beat their wives go on to murder their wives, the question is, how does that figure compare to the people who don't beat their wives? You see, it's not true, that among the whole population of people who don't beat their wives, that 1 out of 1,000 of them murders their wives. That doesn't happen. It's much lower than that. So, in fact, the fact that 1 out of 1,000 people who beat their wives go on to murder their wives actually is an indication that they're in the much smaller subcategory of people who may go on to murder their wives.

The second part of this that makes it a poor argument is simply that, in this particular case, the wife was, in fact, murdered. So, it wasn't the case that we were talking about a hypothetical population of people who had not yet murdered their wives, and then we say 1 out of 1,000 of them will; but instead, we already had a murdered victim. So, that was a silly argument. However, it has the effect of complicating and confusing jurors, and I think this is one of the purposes of statistics used in some jury trials is—how can you confuse the jurors with statistics.

There's one more example I wanted to bring up that came up during the O.J. Simpson trial, but also comes up frequently, and that has to do with DNA evidence.

You've all read that at a murder scene, or a crime scene, there is evidence that is gotten from the blood samples at the scene, and if, for example, the perpetrator of the crime had been cut, and there was blood found at the scene, they run it through tests and they get a DNA description of the blood type in great detail. Then, at the trial, they will present this evidence, and they'll say, "Only 1 in 1 million people have these characteristics in their DNA that we found at the crime scene and, therefore, of course, that's strong evidence that the person who has the corresponding DNA must be guilty."

But we have to be a little careful. Suppose that the following is what the police actually did. Suppose at the crime scene, they found the blood evidence, and they typed the DNA. And then, there was an archive of DNA evidence for all the people in the city; now, we don't have this yet, but suppose there were. And then, they went down the list, and they found somebody in, for example, a city of 10 million people. Suppose they picked out somebody whose blood type matched the DNA at the blood scene, and then they put this person on trial.

Now, you're the jury. How do you evaluate the blood evidence against the defendant? Well, if the DNA type was the reason the defendant was selected as a defendant, then you can't say, "Well, there's only a 1 in 1 million chance that the evidence matches." In fact, you would say the following. You'd say, "If the person were just chosen at random to match the blood type, and if there were 10 people—that's 10 out of 10 million, if it's 1 in a 1 million chance—so if there were 10 people in the city who had that same blood type, and you just chose one at random, in fact, the chance that that person was the guilty party on that evidence is only 1 out of 10."

In any case, I think that these are fascinating issues of where the use of statistical and probabilistic evidence in the criminal justice system can cut both ways, and it requires some really sophisticated thinking. I hope that you enjoyed this discussion of applications of statistics to the law.

Lecture Fourteen
Democracy and Arrow's Impossibility Theorem

Scope:

The most fundamental idea of democracy is that the government responds to the will of the people. Usually, the will of the people is ascertained through voting. An election takes the individual opinions of each voter and assembles those many opinions into one societal decision, the election winner—the will of the people. In this lecture, we will consider an unfortunate and counterintuitive reality about the election process. An election's outcome may have less to do with the voters' preferences about the candidates than with the voting method employed. A method by which the voters' preferences are combined to determine the winner is a means of making a statistical summary of data. We will see that such summaries are fraught with peril. Arrow's Impossibility Theorem proves that every election method has undesirable features.

Outline

I. The most fundamental idea of democracy is that the government responds to the will of the people, but what do we mean by "the will of the people"?

- **A.** An election takes the individual opinions of each voter and assembles those many opinions into one societal decision—the will of the people.
- **B.** We will see that that idea has some serious inherent difficulties.
- **C.** From a statistical point of view, an election is a process of summarizing a set of data.
- **D.** In this lecture, we will consider two counterintuitive realities in the election process.
  1. First, the voters' preferences about the candidates may have less to do with an election's outcome than the actual voting method employed.
  2. Second, every voting method is seriously flawed, and in some sense, the only self-consistent voting method is a dictatorship.

E. These results about the election process provide a cautionary tale about difficulties associated with summaries of data.

II. Elections take the choices of each member of the population and return one societal choice.

A. At first, it seems there is nothing to discuss—an election is held; whoever gets the most votes wins.

B. We will soon see that this method of voting, called *plurality voting*, works great when there are two candidates.

C. When a third candidate wades into the election, problems arise.

III. Let's consider a municipal election among three candidates—two Republicans and one Democrat.

A. In this city, let's assume that Republicans prefer any Republican candidate over any Democratic candidate and Democrats prefer any Democratic candidate over any Republican candidate.

B. Let's suppose that there are a few more Republicans than Democrats in the city. In fact, to ground our discussion, let's assume that there are only 22 voters total—10 Democrats and 12 Republicans.

C. Suppose an important election is held and there are three candidates: Ron Republican, Rick Republican, and Dan Democrat.

D. The voters' preference orders are recorded in the following table:

Rank	**8 Republicans prefer**	**4 Republicans prefer**	**6 Democrats prefer**	**4 Democrats prefer**
1st	Ron	Rick	Dan	Dan
2nd	Rick	Ron	Rick	Ron
3rd	Dan	Dan	Ron	Rick

E. Who should be declared the winner?

IV. There are several reasonable methods for determining the winner of an election. Let's introduce three of them in this situation.

A. Simply count the number of first-place votes.

1. First-place votes are, of course, the votes a person would get if each voter just got to vote for one candidate, which is the usual method.

2. This method of counting the votes is called *plurality voting*.

B. Another method would take into account the second-place preferences, as well.

1. We could let each person vote for two candidates and declare the winner to be the person who gets the most votes.

2. This vote-for-two method avoids electing someone who is viewed as the last-place candidate by a lot of people.

C. The third method we will consider weights the preferences of the voters. That is, each voter gives 2 points to his or her first-place candidate, 1 point to the second-place candidate, and 0 points to the third-place candidate.

1. This method is called a *Borda Count*.

2. Jean-Charles de Borda was a French scientist who was one of the pioneers in the study of voting methods.

V. Let's see how the various candidates fare under these different possible election methods and record the outcomes in the following table.

A. With plurality voting, we simply read across the top row of the preference table. Dan is the winner using plurality voting.

B. With the vote-for-two method, we count how many voters put each candidate in one of the top two rows. Rick wins using the vote-for-two method.

C. To compute the Borda Count, we give 2 points whenever a candidate appears in first place, 1 point for a second-place vote, and 0 for a third-place vote. Ron wins using the Borda Count method.

Voting Method	Ron	Rick	Dan	Winner
Plurality	8	4	10	Dan
Vote-for-Two	16	18	10	Rick
Borda Count	24	22	20	Ron

VI. The voting method determined different winners even though the voters' opinions did not change.

VII. We have not considered all methods. Let's now think about the possibility of a run-off.

- **A.** We can construct an example in which getting more support causes a winning candidate to become a losing candidate.
- **B.** Better is not necessarily better in a run-off method of voting.

VIII. Yet another method of voting is called *pair-wise sequential voting*.

- **A.** The idea is that the candidates are put in some order, Candidate 1, Candidate 2, Candidate 3, Candidate 4, and so on.
- **B.** Then, an election is held between Candidate 1 and Candidate 2.
- **C.** The winner then goes head-to-head against Candidate 3.
- **D.** That winner then goes head-to-head against Candidate 4, and so on, until we are through all the candidates.
- **E.** This voting method can elect someone when every single voter prefers a specific alternative candidate.
- **F.** This method fails to go along with the consensus of all the voters.

IX. Here are three desirable characteristics that we would like to see in a voting system:

- **A.** Go Along with Consensus (*Pareto condition*): If everyone prefers one candidate to another, the lower ranked one should not win. The pair-wise sequential voting system fails this condition.
- **B.** Better Is Better: If more people vote for a winner, that person shouldn't lose. The run-off method fails to have this feature.

C. Irrelevant Is Irrelevant: Suppose a candidate wins the election, then some losing candidate is eliminated; the winner should not then become a loser. Plurality fails this condition.

X. Arrow's Impossibility Theorem proves that no voting method is possible that satisfies the three desirable qualities:

A. Go Along with Consensus

B. Better Is Better

C. Irrelevant Is Irrelevant

D. Arrow's Impossibility Theorem makes us realize that the concept of democratic choice is an intrinsically problematic issue. In the next lecture, we will see that the situation, if possible, gets worse still.

Readings:

Donald G. Saari, *Chaotic Elections! A Mathematician Looks at Voting*.

Questions to Consider:

1. Track meets often score overall team winners by awarding perhaps 5 points for first-place finishes, 3 points for second, and 1 point for third. In other words, a Borda Count method is used, in which different weights are given to the different places. Can you devise a list of voter preferences among three candidates so that under the 2, 1, 0 weighting of votes for first, second, and third choice, Candidate A wins, whereas with the 5, 3, 1 weighting, Candidate B wins?
2. Arrow's Impossibility Theorem is often portrayed as a limitation on democracy. How fundamental an issue do you find it with respect to the concept of a society responding to the will of the people?

Lecture Fourteen—Transcript
Democracy and Arrow's Impossibility Theorem

Welcome back to *Meaning from Data: Statistics Made Clear*. In this lecture, we're going to face the big challenge of democracy, namely that the government responds to the will of the people. That's the basic concept of democracy. But the question is, "What do we mean by the 'will of the people'?" Intuitively, the will of the people simply refers to the idea that we learn the opinions of all of the voters in the country, and we somehow put them together to make a statement about what the will of the whole society is. But we're going to see in this lecture, and the next, that that idea has some very serious inherent difficulties.

Usually the will of the people is ascertained by voting. In an election situation, what an election really does is to take the individual opinions of each of the voters, and assembles them into a societal decision, the will of the people. So, from a statistical point of view then, what an election actually is, is a summary of a set of data—namely, the opinions of the individual voters.

So, in this lecture, we're going to consider two counterintuitive realities about the election process. The first one is that the voters' preferences about the candidates—which you think are going to be the only issue involved—sometimes have less to do with the outcome of an election than the voting method one uses to assemble these votes and make a summary. You see, whenever you have an election, you have to assemble the opinions and choose one winner. So, that choice of method is going to be a primary issue.

The second thing is that we're going to see a very unfortunate reality about voting methods, which is that every voting method has a very serious defect in it—and you'll see what I mean by defects in voting methods—and that these are absolutely unavoidable. So, these results about election processes actually give us a cautionary tale about what to expect when we summarize a collection of data. There are issues, which these things will illustrate.

Let's begin with an election. The goal is to take the opinions, the choices of every single member of the voting population, and return one societal choice. That's the goal of voting. In an election, it accomplishes this by somehow putting those opinions together.

At first, when you think about this, it seems that there's nothing to discuss. What's there to discuss about voting? It's very simple; you ask people what their opinion is, and you count up the ballots, and whoever gets the most votes wins. It's simple. It's a very short lecture today.

But, as a matter of fact, it turns out that we have many different methods of voting. The voting method of simply having everybody vote for one candidate and seeing who gets the most votes is called *plurality voting*. It works completely perfectly if there are two candidates. Because if there are two candidates, whoever gets more votes is the winner, and there's no question. The difficulties arrive when a third candidate arrives on the scene. When the third candidate wades into the election, the waters become very murky, very muddy.

I think the best way to start this whole discussion is with an example. Let's suppose that we have a committee in which there are three candidates. There are two Republican candidates and one Democratic candidate. Let's assume, just for the purposes of this illustration, that on this committee, all the Republicans only prefer Republican candidates over any Democratic candidates; and all the Democrats prefer the Democrat candidate over any Republican candidate. This is just hypothetical, I know, but let's just assume this for the moment.

So, let's suppose that there are a few more Republicans on this committee than there are Democrats. In order to actually specify our situation, let's look at a specific chart here that illustrates the voters' preferences. This table shows that there are 8 Republicans on the committee, and there are 3 candidates. The 3 candidates are Ron, Rick, and Dan. The names are chosen so you can easily remember who the Republicans are and who the Democrats are. The ones beginning with R are the Republican candidates; the one beginning with D is the Democratic candidate.

The Republicans, of course, prefer the Republicans before they prefer the Democrat. The Democrat is in third place for all of these voters. This number up here is the voters on the committee. There are 8 Republicans who prefer Ron first then Rick—of course, Dan the Democrat is last place—and 4 of the Republicans on the committee prefer the Republican Rick over the Republican Ron, and over Dan.

Then there are 6 Democrats on the committee who prefer Dan, then Rick, then Ron. The other 4 Democrats of course prefer Dan—he's the Democrat—then Ron, then Rick. You follow me? So, this chart illustrates the preferences of the 22 people on the committee; 22 because there are 12 Republicans and 10 Democrats.

Now, given these preferences—these opinions of the people on the committee—our question is very simple: Who should be declared the winner of the election among the 3 candidates, Ron, Rick, and Dan? Let's look at the chart and just see.

The simplest way to evaluate the votes would simply be to use the plurality method. This was the method where you simply take the first-place votes of each voter, and you see who gets the most first-place votes, and that person is declared to be the winner. So, in this case, Ron got 8 votes, and then Dan got 10 votes, you see? So, Dan would be the winner because, you see, Rick only got 4 votes.

This is an example where the Republican vote was split. Even though there were 12 Republicans on the committee, since there were 2 Republicans running and 1 Democrat running, the Republican vote was split and, therefore, the Democrat won this election.

Let's see if there are different ways to evaluate the opinions of these committee members to come up with a societal opinion. Maybe Dan is not the best reflection of the true opinions of this committee.

Here's another way. This method has everybody vote for two candidates. Each person has two votes, and they get to vote for two, and then we just see who gets the most votes under those circumstances. In order to count up how many votes each person gets, let's just look at, say, Ron. Ron gets 8 votes here and 4 votes here for a total of 12 votes. All of the Republicans put Ron in first or second place. Then, Ron gets 4 more votes from the Democrats. This group of Democrats, they prefer Ron second over Rick—so the total is 12 + 4 is 16 votes in the vote-for-two method for Ron.

Likewise, we can look at Rick. Rick gets, of course, the 12 Republican votes—they all put all the Republicans in first or second place—but Rick also gets 6 of these Democratic votes because he's in second place for 6, in the opinions of 6 Democrats, so he gets a total of 12 + 6, which is 18 votes. Dan, you notice, picks up no additional votes because the Republicans all put Dan in third place and, therefore, they don't vote for him at all.

Consequently, the effect is that, under this voting scheme, Rick wins the election. Congratulations, Rick. That is a perfectly legitimate way to summarize the opinions of these 22 committee members and say that Rick seems like a good candidate as the winner.

Well, we're not done yet. There's yet a third method for evaluating votes, and that's a way of actually weighting the opinions. In other words, we're giving two points for anybody who gets a first-place vote, one point for getting a second-place vote, and zero points for getting a third-place vote. You follow me? This makes sense, that the person who's in first place should get more benefit from being the first choice rather than the second choice.

Look what happens here. In this way of getting a weighted sum of the votes—called the *Borda Count*—the opinions change yet again. Let's look at, say, Ron's votes. He gets 8 first-place votes, so that's 8 × 2, which is 16; plus he gets 4 second-place votes here, so that's 4 more, for a total of 20; and then he gets an additional 4 votes here for a total of 24. There is his score, 24.

In this case, though, with Rick, let's see what happens to him. Rick gets only 4 first-place votes in this column, so that's 4 × 2, which is 8; plus 8 second-place votes here, so that's 8 more—8 + 8 is 16; plus 6 votes here—16 + 6 is 22. Rick gets 22 in the Borda Count and, counting the same way, Dan gets 20 votes. In the Borda Count method then, Ron wins the election. Congratulations, Ron.

Now, notice what we have here. This is really quite interesting because what we have shown is, with the same preferences—we didn't change the opinions of the people; the committee's preferences were given here on this list—and yet, we just used three different reasonable methods for getting their votes to make one decision, and we got three different answers. All three of the candidates, from one point of view, should be viewed as the winner.

Well, this is a troubling kind of issue. It's an example of a case where the method of voting has more to do with selecting the winner of the election than the preferences of the people.

It turns out that a mathematician by the name of Donald Saari has proved a theorem that says you can devise voting preference schemes, like this chart up here, so that if you have any number of candidates—for example, you have 10 candidates—you can make it so that the plurality winner is one of the candidates. If you have a vote-for-two method, you get a different winner; if you have a vote-for-three method, you get a different winner; if you have a vote-for-four method, you have a different winner; all the way down to 9 different winners. Then, if you have the Borda Count, you get the 10th winner. Every single one of those 10 candidates would win under these different voting schemes. It's very interesting.

By the way, the Borda Count is named after a French scientist, Jean-Charles de Borda, who was one of the pioneers in the study of voting methods.

You may say the problem with this is that we didn't use an obvious method, which is the method of taking a run-off. Let's proceed then and investigate the question of run-offs because maybe run-off elections would be the solutions to our problems here. Why don't we just have a run-off? Because the run-off would have solved the problem here of electing a Republican, at least.

The problem with the plurality election in this previous example is that Dan, although he won the plurality vote, really would have lost the election to either Rick or to Ron. In a way, we could argue that Dan should certainly not have been the winner, even though he was the winner of the plurality method. But in a run-off, if there were a run-off between two candidates, the Republican would have won because all the Republicans would have voted for the Republican candidate.

But let's see a little difficulty with the concept of a run-off election. Here's an example where we have a different election situation. We have 31 Republicans and 25 Democrats in this voting population—so think of it as a big committee of 56 people—and there are 5 people running for office. Again, we have the 2 Democrats whose names start with D, and 3 Republicans whose names start with R. Suppose that the Republican vote is rather evenly split among their 3 candidates, and the Democratic vote is split rather evenly between their 2 candidates.

Look what happens if we have a run-off. We take the top two vote-getters, which are Dan and David; we have a run-off between Dan and David, and David or Dan wins. We have a Democrat winning this contest. Let's say David would win the election. Of course, it would depend on how the Republicans voted in the run-off, but let's suppose that David was the eventual winner under this run-off scenario.

Look what happens if David does a little better job among his own constituency, among the Democrats. He does a better job; he gets more votes among the Democrats in the first election. In this case, the 25 Democrats vote 15 for David and 10 for Dan. In other words, David did better in that first election. Let's see what happens.

Well, it's a run-off, remember? Who are the contestants in the run-off? The run-off contestants are David and Rhonda. In the run-off, it's very clear who's going to win the run-off election, right? Namely, Rhonda. So, Rhonda wins the run-off election because there are 31 Republicans who are going to vote for her, and only 25 Democrats who will vote for David. So, the consequence of David having done better in the preliminary trial was that he did worse overall; he lost the election. An interesting feature of the run-off kind of election, that doing better can lead to a worse result, and it did in this example.

Now, here's another method of voting that is called *pair-wise sequential voting*. This method of voting is a method by which you simply take two of the candidates, and you pair them against each other. Whoever wins then goes against the next candidate; they go against each other; and then you get to the last candidate; they go against each other; and whoever wins at the end wins the election.

Let's see what happens if we have this election situation. Suppose we have 3 voters, a small committee. They're going to make a decision among 4 different candidates to appoint: A, B, C, and D. Maybe it's a committee that's appointing an officer in the company, and this 3-person committee is choosing among 4 candidates. The 4 candidates are A, B, C, and D, and they decide to use the sequential pair-wise voting method. So, let's just see what happens.

They first pair A against B. You see this could very well happen. When you're trying to sift through candidates for this position, you may say, "Well, let's look at these two. Which one of these two is better?" There's a discussion, and an opinion is given.

Well, so, let's look at A against B. Voter 1 thinks that A is better than B. That's what this chart means. Voter 2 also thinks A is better than B. Voter 3 actually thinks B is better than A; but since 2 of them think A is better than B, A would win that first pair-wise contest. So, then, we proceed and compare A to C. You can imagine this in a committee, right? You just go through the people, and you say, "This one's better." You throw out B and go on to C.

Now let's look at A. A is still in the contest, and now A is being compared to C. So, A goes against C. Voter 1 thinks A is better than C. Voter 2 thinks C is better than A. Likewise, Voter 3 believes that C is better than A. So, C wins the pair-wise contest, and we proceed to see that now C beats A in this pair-wise contest.

Now C is the remaining candidate. We pair C to D, and look what happens. D is better than C here; C is better than D here; but D is better than C here. So, two of the people think D is better than C and, therefore, D beats C. So, D would be the winner and would be selected and appointed to the board, or to the position.

Well, there's only one slight problem with this—a rather major problem—and that is everyone likes B better than D. Look at this. Voter 1 thinks B is a better candidate than D; Voter 2 thinks B is a better candidate than D; Voter 3 thinks B is a better candidate than D. So, this selection method has chosen a candidate as the winner where all people in the voting population think that one specific other candidate is superior. Isn't that a strange thing? Well, that is a rather unfortunate kind of a quality for a voting system to have, that it could choose a candidate when there is a consensus in the population that somebody else is the better candidate.

Actually, we've been discussing several kinds of properties that we might want in a voting system. One of them is consensus. If there is a consensus, we should go along with it. So, one desirable property of a voting system would be that we want to Go Along with Consensus. Unfortunately, the pair-wise sequential voting system fails that method.

Another quality that we would really like to have is the quality that Better is Better. That means, if you do better, if more people want you and vote for you, prefer you in a higher position, then that should help you in the outcome, not hurt you. But, remember, that we saw in the case of the run-off elections that run-offs fail this property. Remember, in the run-off, in the first instance where the vote was split between the two Democratic candidates Dave and Dan, Dave went on to win the election because both of the candidates in the run-off were Democrats. Whereas, when Dave did better in the preliminary vote, in the first vote, he did better and, consequently, the second-place person was a Republican. So, consequently, in the run-off, he was running against a Republican and, therefore, the Republican won, where in our scenario, there were more Republicans than Democrats in the voting population. So, the run-off election fails this property that Better is Better, and that's a desirable property that we would want a voting system to have.

The third desirable property that we would like a voting system to have is the idea that Irrelevant is Irrelevant. What I mean by that is, suppose that you have an election, and there are several candidates, and then you throw out one of the losing candidates. Let's say you have a method of voting; you have an election; and there's a winner. Then you discover after the election that one of the candidates maybe didn't qualify to be in the election, and you throw that person out; they lost anyway.

Suppose it has the effect of making somebody else the winner. You see, that doesn't seem like a good quality for an election system to have. Yet, plurality voting has exactly that property, right? If you threw out a losing candidate, it can change the order of the winner and the loser.

This isn't just some abstract issue, by the way. In the year 2000, in the presidential election, there were three candidates: Bush, Gore, and Nader. If Nader had not been in the election, Gore would have been the victor, right? Because Gore against Bush would have won, and, certainly, Gore against Nader would have won. Gore would have won the election. So, the point is that plurality voting has the property that if you throw out a losing candidate, it may change the order of who wins the election.

Of course, presidential elections are not decided using plurality voting of the popular vote; however, the 2000 election does give an example where eliminating a losing candidate would have changed the outcome.

You might wonder why it is that I keep talking about all these methods that don't work. Why don't I just come out and tell you the one that has these nice, desirable features that we've identified? The reason I don't do that—as you've probably guessed at this point—is that it's impossible. There is no voting system that allows us to have all three of those properties in place. This is Arrow's Impossibility Theorem.

What is a voting system? A voting system is a method by which you take the preferences of all the people in the voting population and from them, assemble a winner or an ordered list of the societal preferences. What Arrow's Impossibility Theorem says is that there is no system for summarizing the opinions of people that satisfies all three of these conditions: Go Along with Consensus—which, by the way, is called the Pareto condition; Better is Better, meaning that if a candidate does better, it should advance him or her in the outcome of the societal summary list, rather than decreasing that person's result; and Irrelevant is Irrelevant, meaning if you throw out a losing candidate, it should not have the effect of changing the order of the other candidates.

There is only one system that has all three of these, and that is a dictatorship. If you just say there is one voter in the country, and whatever it is that that voter says goes, that does, in fact, satisfy all three of these conditions.

One of the implications for Arrow's Impossibility Theorem is on the plus side—we've got to have at least a little ray of sunshine in this lecture—and that is that we will never have to worry that the bowl championship series method of evaluating the college football teams will be settled. We'll always be able to argue about them because we know that any method that they choose—no matter how many computers are involved, or opinions of sports writers. No matter what method we use—one of those three conditions is not going to be met, and those issues will arise. And believe me, people will say, "How could you possibly choose a voting method that fails to Go Along with Consensus?" Or where, if you throw out this losing team, it changes the order. They will definitely bring those issues up. That is a very cheerful thought that we can look forward to, debates in the future.

In the next lecture, we will continue this theme of the discussion of voting methods and, if possible, we will discover that, in fact, the situation gets even worse. This is not a happy lecture. Why am I giving these lectures? It gets worse, but it does have the implication that not only is it worse, but that you can apply the same analogous situations in many different statistical applications that have nothing to do with voting and realize that statistical summaries entail significant problems.

I'll look forward to seeing you next time.

Lecture Fifteen
Election Problems and Engine Failure

Scope:

This lecture begins as a continuation of the previous lecture by looking at some famous real elections in which we might wonder whether the will of the people prevailed. The challenge of choosing an election winner can be thought of as taking voters' rank orderings of the candidates and returning a societal rank ordering. An analogous and mathematically similar situation occurs in a totally different setting. Suppose we are trying to determine which type of engine lasts longest among several competing versions. One statistical strategy for making such a selection is to run several of each type of engine until they expire, then to put the experimental results in order of longevity. Of course, all is well if one type of engine always lasts longer than the others. However, in reality, the correspondence of lifetime-to-type might be less consistent. Combining those data into one choice of engine with longest expected life incurs the same paradoxical difficulties that we previously encountered in our analysis of elections.

Outline

I. In this lecture, we begin with a case study involving real data from an important election.

A. Here is a chart that summarizes the voting preferences of the population, giving for each candidate the percentage of the population that ranked the candidate first, second, and so on.

Candidate	% of 1st	% of 2nd	% of 1+2	% of 3	% of 1+2+3
A	40	14	54	16	70
B	13	46	59	33	92
C	18	18	36	3	39
D	29	22	51	48	99
Total %	100	100	200	100	300

B. What follows is a summary of how well each candidate would have done under each voting scheme.

1. In plurality voting, A would win.

2. In the vote-for-two scheme, B would win.
3. Using a run-off, D would win.
4. Using the Borda Count method, D would win.

II. What candidate do you think best represents the will of the people?

A. Many people feel that Candidate D is the best choice.

B. Before telling you who actually won this election, let me tell you that this election was a presidential election; thus, there is one more column to be added to the table, namely, the Electoral College votes.

1. In the Electoral College, Candidate A won handily with 180 electoral votes.
2. Interestingly Candidate C, who was not a contender in any of the other schemes presented, actually received the second highest number of electoral votes.
3. Perhaps now is the time to tell you what election this was: It was the presidential election of 1860.
4. Candidate A is Abraham Lincoln, B is Bell, C is Breckinridge, and D is Douglas.

C. The percentages of people who rated the candidates in second and third places are estimates obtained from historians of the Civil War.

D. It is intriguing to think about the consequences of the statistical issue of how to summarize the data of the voters' opinions.

Candidate	**Plurality**	**Vote-for-Two**	**Run-Off**	**Borda Count**	**Electoral College**
A	40%	27%	46%	164	180
B	13%	30%		165	39
C	18%	18%		92	72
D	29%	25%	54%	179	12
Winner	A	B	D	D	A

III. We have seen lots of bad news about election problems. Now, if possible, it gets worse. Next, we will discuss the *Condorcet Paradox*.

A. Let's look at an example of an election among three candidates, A, B, and C.

B. The following chart summarizes the views of the 30 voters about these candidates.

Rank	10 Voters	10 Voters	10 Voters
1st	A	B	C
2nd	B	C	A
3rd	C	A	B

C. For every candidate, two-thirds of the people have a specific alternative candidate whom they prefer.

D. We can produce even worse cases—with 10 candidates, for example, when no matter who is declared the winner, there is a specific alternative candidate who is preferred by 90% of the voters.

IV. Many people feel that if there is a specific candidate who would beat every other candidate in a head-to-head contest, then that candidate, the *Condorcet winner*, should be declared the victor.

A. Marie Jean Antoine Nicolas Caritat, marquis de Condorcet, wanted to point out a weakness of the Borda Count method.

1. He did so by producing the following famous voters' profile:

Rank	30	10	10	1	29	1
1st	A	B	C	A	B	C
2nd	B	C	A	C	A	B
3rd	C	A	B	B	C	A

2. B wins with the Borda Count.

3. But A is the Condorcet winner.

B. Thus, the Borda Count method does not always select the Condorcet winner.

C. However, in a sort of voting theory double-reverse, recently, Donald Saari has pointed out an interesting further analysis of this old example.

1. If we first erased collections of voters whose votes canceled one another, then B should win.
2. Perhaps the Borda Count method chose the right winner.

20	**0**	**0**	**0**	**28**	**0**
A	B	C	A	B	C
B	C	A	C	A	B
C	A	B	B	C	A

D. This example shows again the subtleties of summarizing data meaningfully.

V. The voting method was decisive in choosing the location of the 2000 Olympic Games.

A. The method used was plurality voting, in which the bottom-ranked choice is systematically eliminated and the remaining cities are voted upon.

B. Here's what happened:

1. Starting with five cities, in the first three contests, Beijing was the victor.
2. In the final vote, Sydney won.

C. Sydney, not Beijing, hosted the 2000 Olympics.

City	**1st Vote**	**2nd Vote**	**3rd Vote**	**4th Vote**
Beijing	32	37	40	43
Sydney	30	30	37	45–**Win**
Manchester	11	13	11	
Berlin	9	9		
Istanbul	7			

VI. These voting paradoxes are examples of trying to summarize a set of data that has reflections beyond voting.

A. Another example: How can we tell which of three engine brands lasts longest?

B. There is variability among the engines produced by each of the three contenders, of course, so we can't just run one engine from each company and see which lasts longest.

C. Here is a method called the *Kruskal-Wallis test*.

1. We take several engines, say five for illustrative purposes, from each company.
2. We run all the engines until they fail.
3. We score the engines 1, 2, 3, …, 15 based on how long they lasted, with the longest lasting scored 1.
4. We add up the scores for each company's engines.
5. The lowest number wins.

	Eng. 1	**Eng. 2**	**Eng. 3**	**Eng. 4**	**Eng. 5**
A's Time to Failure	1,137	993	472	256	207
B's Time to Failure	1,088	659	493	259	238
C's Time to Failure	756	669	372	240	202

	Eng. 1	**Eng. 2**	**Eng. 3**	**Eng. 4**	**Eng. 5**	**Total**
A's Failure Order	1	3	8	11	14	**37**
B's Failure Order	2	6	7	10	13	38
C's Failure Order	4	5	9	12	15	45

D. The Kruskal-Wallis technique has defects similar to those we saw in voting methods.

1. In this example, suppose we eliminate C's engines.
2. Then, the recomputed Kruskal-Wallis test would indicate that B's engines are superior.

	Eng. 1	Eng. 2	Eng. 3	Eng. 4	Eng. 5	Total
A's Failure Order	1	3	6	8	10	28
B's Failure Order	2	4	5	7	9	**27**

E. All the examples in the last two lectures suggest that summaries of complex situations require contextual arguments to decide among them.

1. Statistical and logical analysis can help a great deal in choosing which arguments to find most persuasive and which systems to use in which settings.
2. Voting theory is an intriguing topic for further study.

Readings:

Donald G. Saari, *Chaotic Elections! A Mathematician Looks at Voting.*

Questions to Consider:

1. Strategic voting is encouraged when some people are better off voting for someone they don't really want in order to elect the ones they do want. In which of these voting methods—the Borda Count, run-offs, plurality, vote-for-two—is strategic voting encouraged? Are some methods more susceptible to strategic voting than others?
2. A voting method we did not discuss much is approval voting, in which each voter can vote for as many or as few candidates as the voter finds acceptable. Many people find this an attractive system. What are the pros and cons of this system?

Lecture Fifteen—Transcript
Election Problems and Engine Failure

Welcome back. In the previous lecture, we discussed voting schemes and the problems with them, and this is a continuation of that discussion about voting problems. Last lecture, we introduced several different methods for taking the opinions of the voters from a population and choosing the winner from those opinions. Among them were plurality voting, the common one, where you just see what candidate got the most first-place votes. Then, we had vote-for-two method, where every voter could vote for 2 candidates. We had a method called the Borda Count, where we weighted the voters. If you had 3 candidates, the voter could give 2 points to their first choice, 1 point to the second choice, and 0 to their third choice; the numbers are then added up. Then, we had yet another method, namely a run-off method.

We saw some problems with all of these different methods of assembling the opinions of the voting population and assembling them into a societal decision. But, in real life, we have to actually make a decision; we have to decide who the winner is in a contest.

So, in this lecture, we're going to start out with a case study, where we look at the preferences of the people and ask the question of "Who is the people's choice?" This example actually is a real election—it was real data from an important election—so it will be interesting for you to see it and make your own decision about what you think really would be called the people's choice.

Let's look at this chart here. This chart is formatted in a little bit different way from the previous charts that indicate how the population prefers their candidates. In this chart, this is the way to read it. There are 4 candidates: A, B, C, and D. The first column here tells us the percentage of the voting population who viewed each of the candidates in first place. In other words, this 40 means that 40% of the voting population viewed Candidate A as their first choice; 13% viewed Candidate B as their first choice; 18% viewed Candidate C as their first choice; 29% viewed Candidate D as their first choice. So, in particular, A got the most first-place votes.

The second column indicates what percentage of the population considered the individual candidates their second choice: 14% for A; 46% for B; 18% for C; and 22% for D. In other words, a lot of people thought Candidate B was a good second-place person. Not too many—only 13%—viewed that candidate as their first choice, but a lot of people—almost half of the whole population—thought of that candidate as their second choice. Then, this column simply adds up these numbers of the first and second place. So, it really comes out of 200 because it's not a percentage anymore.

This third column is similar to this column. This column is the percentage of people in the population who considered the candidate to be their third-choice candidate out of the 4. So, 16% viewed Candidate A as their third choice; 33% of the population thought of Candidate B as the third choice; 3% for C; and 48% for D.

Then, this last column represents the total of the percentages—those first place, second place, and third place. So this is out of 300. Notice in this third column, this says that 99% of the people viewed Candidate D in one of the first 3 places; and 92% of the voting population viewed Candidate B in one of the first 3 places. So, you understand the meaning of this chart; it's capturing the opinions of the people.

Now let's go to the trouble of trying to think, looking at this chart, what do you think is the will of the people? Looking at these data, what would you say that people really want? Which of the candidates do you think would be described as the will of the people?

Let's first of all look at different ways that you might evaluate that societal decision. One is that you could simply look at the first column, which is the plurality voting system. Namely, you just look and see how many people want each of the 4 candidates in first place. In that method of evaluating the societal decision, we see that Candidate A wins because 40% of the people viewed Candidate A as their first choice.

Now, there are different methods, as we saw in the last lecture, for evaluating the societal decision. Another way would be to say, let's let everybody vote for 2 candidates. If we did that, we have the answer here in this column. Each number in this column tells us the percentage of the voters who would give one of their two votes to that candidate. So we see that Candidate B would win using the vote-for-two method, since B received more votes than any other candidate.

Let's go on to the next possible way of evaluating the winner of the election, and that would be to hold a run-off. In other words, we'd look at the plurality first-place and second-place candidates. That would be Candidate A and Candidate D. Then, we would have a run-off between these 2 candidates. This chart really doesn't speak to the entire question. You can't get who the winner would be from those data; but other information told us that, in fact, in a run-off election, it would have been Candidate D who won the election. In the run-off method of evaluating it, D would be the winner.

Remember the Borda Count, which is the count where we weight the votes. In other words, in this case, since there are 4 candidates, each voter could give 3 points to his or her first-place candidate, 2 to the second-place, 1 to the third-place, and 0 to the fourth-place candidate. Then adding up those numbers for each voter in the population, we would get a number. In the Borda Count method, we see that the numbers are 164, 164, and 179—so the Borda Count winner is Candidate D.

I'd like you to think about this yourselves and say who you think should win. You can see that we have a variety. We have A, B, and D as all potential winners for this election, and it's a little bit hard to know which one you should choose. As I've talked about this chart with people and had students and others groups think about it, generally, most people view that Candidate D would be the better representative of societal choice.

First of all, in a head-to-head election between A and D, Candidate D wins by 54% to 46% of the vote. It's true that Candidate A won the plurality vote, had more first-place votes, but in the Borda Count method, Candidate D won. And also, 99% of the voters in the population at least put Candidate D in the top 3; whereas, there were 30% of the people for Candidate A who didn't view Candidate A as an acceptable choice—they put him in last place.

Well, let me tell you now that actually there's one more column that we should add to these methods of evaluating votes, and that is the Electoral College. It turns out that these data have to do with a Presidential election, and the Electoral College had this result. In fact, in the Electoral College, the winner of the election was Candidate A, with a rather a strong victory—180 votes in the Electoral College. Notice, interestingly, that Candidate C, who was really not one of the contenders, was the second-place winner in the Electoral College.

This is interesting, and one reason that this is actually a very interesting result is that this was an extremely important election. This was the Presidential election of 1860. Candidate A was Abraham Lincoln; Candidate B was Bell; Candidate C was Breckinridge; and Candidate D was Douglas. We can easily imagine that if a different method of choosing the societal preferences were employed, the entire history of the United States would have been extremely altered in an extremely major way. It may well have been that the Civil War would not have been fought if Douglas had been President. Of course, we don't know that. The point is that these concepts of taking statistical summaries and giving an answer have extremely serious consequences in real life.

I promised in the last lecture that, if anything, matters about voting would get worse. So, I'm about to deliver on that promise by looking at this next example, which is the *Condorcet Paradox*. Marie Jean Antoine Nicolas Caritat, the Marquis de Condorcet, was a French scientist who was also an early pioneer of voting theory, just like Borda was. One of the things that he pointed out was this very interesting paradox about voting.

Suppose that we have the following voting preferences among a population. In this case, we have 30 voters. And the 30 voters prefer the 3 candidates—A, B, C—in the following way: 10 voters prefer A as first place, B as second place, and C as third place. The order of preferences for these 10 voters is A, B, and C. For these 10 voters, it's B, C, A, just permuting A, B, C, putting the A on the bottom. Then, we permute it once again to get C, A, B. So, you see that these things are just a permutation, meaning that the A is moved down to the bottom, and then the B is moved down to the bottom to get this column. You see that? So, in other words, 10 people want A; 10 people want B; and 10 people want C as their first-place vote.

Who should win this election? Well, this election has a problem. Suppose that we were to select a winner, for example, Candidate A. We have to choose somebody—A, B, or C—so suppose we chose Candidate A as a winner. Well, in a way, Candidate A would be a very bad choice for a winner because of the fact that 2/3 of the voters prefer C to A. Look at this. You see, these 10 voters prefer C; these

10 voters prefer C. So 2/3 of the population would prefer C. So, that seems like a poor choice for an election, to come out with a winner where 2/3 of the people want somebody else.

Maybe B's a better candidate, so let's try B. B is a bad choice because 20 people prefer A to B. You see, these 10 prefer A to B; these 10 prefer A to B, so B is a bad choice. In fact, here's another one, C. Let's try C as the possible winner. Well, C is a bad choice because, look, these 10 people prefer B to C, and these 10 people prefer B to C. No matter what winner we declare, the problem is that 2/3 of the voting population will prefer a specific alternative candidate.

So, this is the Condorcet Paradox. Well, in fact, it gets yet worse. This was a case of 3 candidates. Suppose you had, for example, 10 candidates—and it can work with any number of candidates—and you labeled them A, B, C, D, E, F, G, H, I, J. You could have some voters prefer them in that order, and then you could do the same permutation business—that is, putting the A at the bottom and everybody slides up one—and you have the same number of voters prefer them in that order. Then you could take B off and put it at the bottom, so it would end with A, B, and the same number of people want them. Then put C at the bottom, and so on.

So, if your voting population were divided into 10 equal groups so that everyone preferred the candidates in this different order, no matter what candidate you chose, there would be an alternative candidate, which 90% of the voting population would prefer. There's no choice. No matter which candidate you preferred, you have that kind of outcome. This is the Condorcet Paradox.

In the early discussions of voting theory, two of the pioneers in the discussion of these issues about voting theory, as I've already said, were Borda and Condorcet. There was an issue between the two gentlemen, and that had to do with a concept called the *Condorcet winner*. The Condorcet winner is a candidate who could beat any of the other candidates in a one-to-one contest. So, if there is an individual candidate who could beat every single one of the alternative candidates, pair-wise, then that person might be viewed as an appropriate choice for a winner. Many people think that if there is a Condorcet winner—which it is not necessarily the case that there is always such a person—but if there is a Condorcet winner—somebody who will beat every other candidate in a head-to-head contest—then you might want to think that the election should choose that person.

Notice, by the way, that in real life, we do not select such people in our actual voting practice. For example, in the Electoral College method of choosing Presidential winners, we can look at several years in which that result did not happen. In the 1992 election, where the three candidates were George Bush, Sr.; Clinton; and Perot, it was commonly viewed that Perot had taken voters away from Bush, and that Bush would have beaten Clinton in a head-to-head contest; or Bush would have beaten Perot in a head-to-head contest. So, Bush would have been the Condorcet winner of the popular opinion.

Likewise, in the year 2000 election, Gore would have beat George Bush, Jr., in a head-to-head contest and, of course, would have beat Nader in a head-to-head contest. So, Gore would have been the Condorcet winner in that contest. So, the Electoral College method doesn't choose the Condorcet winner, necessarily.

Well, Borda and Condorcet were discussing these kinds of issues, and they were proposing different voting methods. One of the voting methods was the Borda Count that we've already seen, where you weight the candidates. Using the Borda Count method, a voter gives more points for their first-choice candidate, fewer for the second choice, fewer for the third choice, and so on, and then you add up those points, and it gives a weighted value of the voting. That's the Borda Count.

There was a question about whether the Borda Count was a good method of voting. Well, Condorcet—and I think there was a little bit of rivalry between these two people—devised the following chart. This is a chart of potential voting preferences among three candidates, A, B, and C. Let me just tell you what the chart involves. Thirty people say A over B over C; ten people say B over C over A; and so on. That's the way to read this chart.

It has the property that if you do the Borda Count, you see that B actually wins the Borda Count with 109. I won't go through the arithmetic to show that, but B is the winner. The reason Condorcet created this example is that, in fact, A is the Condorcet winner. In other words, A beats B in a head-to-head contest, 41 to 40; A beats C head-to-head, 60 to 21. You can go through the numbers and see that, in fact, those numbers are correct.

So, the Borda Count method did not select the Condorcet winner. This is a very famous example—centuries old—but it has an interesting modern sequel to it. Namely, remember in the Condorcet Paradox where you had A, B, C, and then it was permuted B, C, A, and permuted C, A, B? What we really want to say about that is that everyone is equal, right? If you have those three situations, the voters are saying there's a tie.

A modern development by the mathematician Donald Saari is that he took this classical example and said, "Let's remove all of those cycles where you could view those as a wash, as equal." If you erase those cycles, you get this chart, where you see that, in fact, B should be the winner over A, 28 to 20. Because when you erase all the cycles, which you could view as equal, it turns out that B is the winner. So, there's another way to look at this example, in which the concept of B being a better choice than the Condorcet winner, in this example, illustrates that maybe the concept of the Condorcet winner as the obvious winner in an election is not necessarily correct.

Isn't this interesting? There are just so many fascinating things with these elections. But, I wanted to show you that the application of these kinds of paradoxes lapses over into realms that don't have to do with elections.

So, we have seen in practice that there are challenges with how to choose the winner of an election; but, of course, people have to actually choose the winner. One example of a method for choosing the winner occurs in the case of trying to select the locations for the Olympic Games. In trying to select the winner of the location in their bids for the year 2000 Olympic Games, there were five major candidates near the end of the process who had submitted requests to be the holder of the 2000 Olympic Games. The method of voting at that time, for choosing the winner, was to do the following. They started with all of the contenders, and each person who was voting could vote for one city as the winner. So, each person on the Olympic committee would give a vote to one of those five cities.

In the first vote, Beijing was the winner of that election, 32 to 30 to 11 to 9 to 7. Then, the method was that they eliminated the last-place city, and then voted again on the remaining four cities. The election then changed to this one. In the second vote, 37 people were for Beijing—again, Beijing was the winner—Sydney was second; Manchester was third; and Berlin was fourth. Then, since Berlin had the fewest votes, it was eliminated, and there were three contenders left. In the third vote, Beijing—again, now for the first, second, third time was the winner, 40 to 37 to 11. Manchester was eliminated from the voting; and then in the last vote, Beijing lost the election to Sydney, and the 2000 Olympics were held in Sydney.

This was an example of an election procedure where Beijing, although winning the first three of those votes, lost the last vote and, in fact, did not hold the Olympics. That was an interesting example where we have to make a choice about how to select a winner, and this particular method had that result.

The same kind of paradoxes that we have seen associated with elections can occur in other kinds of statistical strategies. Here's an example of a strategy by which you're trying to evaluate the quality of an engine. A good way to do this would be to take a collection of engines and run the engines until they fail. You see how many days go by before the engine quits, and then you record that number. After you record that number of failures, then you have some data.

You have several different engines of the same manufacturer. You're trying to decide whether an engine made by A or B or C is the best of those three manufacturers. Of course, an individual engine from a certain manufacturer may last a different amount of time from a different one, and so you choose several engines to do this experiment on. You take five engines from each of these companies—five from A, 5 from B, five from C—and you just run the engines until they fail. This engine lasted 1,137 days; this one, 993 days; and so on.

Now the question is to look at these data about the time to failure of the engine and decide which company to buy your engine from. Maybe you're a manufacturer of something that uses engines and you want to know what company to buy from.

You have these data. A method for evaluating this is called the *Kruskal-Wallis test*, which is, instead of looking at the individual numbers, we simply put the numbers in order. That is, we say which engine lasted the longest, and we call that Number One. Then, which engine lasted the second-longest? It was B's engine here, which lasted 1,088 days, so it's Number Two. Then the third longest-running engine here lasted 993 days, and that's Engine Three. Then the fourth one was down here, C's engine, which lasted 756 days—

that is Number Four. Number Five lasted 669 days, here. Number six is here, and so on.

Do you see what I mean? We take all of these days, and we just ignore the big numbers, but we just put them in order and see the order in which the engines failed. Then, all we do is add up the ordinal number of these engines for Company A, B, and C, and we're looking for the company that has the smallest sum of those ordinal numbers because the smallest would mean that those were the ones that were closest to the longest-lasting.

Well, if we do this in this case, we see that the sum of the ordinal numbers for the failure of A's engines was 37; B's 38; and C's 45. So, Company A looks like the company to buy engines from. But notice what happens if we do the process of eliminating the engines from Company C. Company C was the company whose engines we weren't going to buy because, look, their ordinal numbers were quite a bit higher than the other company's—so we certainly don't want them. So, you might think we should throw out those engines from our experiment and just look at these two rows of data.

Look what happens. When we look at the two rows of data, if we throw out these engines and then look at the ordinal numbers of the remaining engines, what happens? It switches. Instead of Company A being the company that you'd buy your engines from, you now conclude that it's B who makes the better engines. So, you see that that's the same paradox that we previously called the Irrelevant-is-Irrelevant paradox that arises in this statistical evaluation of the quality of engines.

All of the examples in the last two lectures force us to ask the question: Is one analysis or one voting method correct and the other ones wrong? The answer, the take-home lesson, is that summaries of complex situations require contextual arguments to decide among them. The validity or the persuasiveness of data summaries is truly a matter of opinion and judgment. Statistical and logical analyses can help us in dealing with choosing arguments to find which are more persuasive and which systems to use in various settings. But the point of these lectures is to say it's up to us to really think carefully and evaluate the context of the statistical information. But statistics is a big help to clarify what those issues really are.

I'll look forward to seeing you next time.

Lecture Sixteen
Sports—Who's Best of All Time?

Scope:

Analyzing sports statistics is a sport of its own. We record statistics about the performances of individuals and teams, then use those data to bolster our arguments about sports prowess. In this lecture, we will examine a couple of statistical questions that illustrate principles applicable well beyond their sporting origins. We will discuss the question: "Who is the best hitter in baseball history?" This question immediately presents statistical challenges that concern comparisons of performance in different eras and different circumstances. Next, we will consider the question of streaks. Do athletes enter "the zone" and have a "hot hand" for periods of time? What is the correct interpretation of slumps and streaks? Such questions force us to confront our understanding—or misunderstanding—of what to expect from randomness. The question, "What is random?" lies at the heart of the issue of streaks and slumps.

Outline

I. Statistics about sports are fun. They also help us to understand sports and appreciate and evaluate the success of individuals and teams.

- **A.** Statistics are useful in discussing relative performance and in making managerial decisions.
- **B.** Sports statistics form good illustrations of key statistical ideas.

II. Who was the best hitter in the history of major league baseball?

- **A.** This question forces us to clarify the relationship between what is measured and what quality we are trying to describe.
  1. The batting average is basically defined as the number of times the batter gets a hit divided by the number of times he is at bat.
  2. We'll also assume that we mean the batting average over a single season.

3. And we'll ignore those players who didn't have many at-bats over the course of the season.
4. All of the 18 highest batting averages in a season in the history of professional major league baseball occurred before 1942. Why?

B. We might suspect, for example, that batting averages in general, that is, the averages of all major league players, were higher in the earlier years of baseball than in recent years.

C. Comparing the 1920 histogram to the histogram of batting averages in 2000, we find that the center, the mean, of the two is about the same (about .265), but the 1920 histogram is more spread out.

1. The standard deviation of the 1920 data set is .050, larger than the standard deviation of the 2000 data set, which is .038.
2. Recall that we expect 68% of the data to be within 1 standard deviation of the mean and 95%, within 2 standard deviations of the mean.

D. Comparing standard deviations away from the mean is a method of normalizing the comparisons over the different eras. In a sense, it measures how well a person performed relative to his contemporaries.

1. The number of standard deviations that a batting average is away from the mean is not necessarily an integer. Every batting average in any given year could be described by how many standard deviations it is above or below the mean.
2. Recall that the number of standard deviations away from the mean is the z-score.

E. One way to measure across eras would be to measure how many standard deviations above the mean a batter's average is.

1. For example, given that Joe Jackson of the Chicago White Sox had a batting average in 1920 that is 2.36 standard deviations above the mean for that year and that Moises Alou of the Houston Astros had a batting average in 2000 that is 2.31 standard deviations above the mean for the year 2000, we might consider those two players about equally good batters.
2. They are about equally extreme outliers.

F. We could list the 10 batters whose batting averages were the greatest number of standard deviations above the mean for their years and declare a winner on that basis.

G. Stephen Jay Gould opines that pitching, fielding, and batting have all gotten better over the years, and their approach to human limits of perfection accounts for the lower standard deviation. Making an interpretation such as Gould's can be helpful in understanding the data.

III. Another complication to the question of who is the best hitter in baseball history is the fact that doubles, triples, and homeruns are more valuable than singles. Likewise, walks are not recognized, although extremely important.

A. Other measures of offensive prowess, such as slugging percentage and on-base percentage, can be used.

B. People come up with various formulas combining these sorts of raw statistics, attempting to get a measure that is highly correlated with helping the team to win games.

IV. Our second statistical issue from sports is the question of the "hot hand."

A. A commentator is often heard to say, "This player is on a streak. He can't seem to miss."

B. Is there really such a thing as a "hot hand," meaning that the player is better for a period of time? Or are the streaks (which are real) accounted for by random luck alone?

C. Suppose I flip a coin.

1. If I flip lots of coins over and again, there will be, from time to time, long streaks of heads.

2. We would not ascribe the streaks of H's and T's in the flipped coins to some property that has changed in the coin for that time.

D. Likewise, if we have an NBA player whose lifetime percentage of making a shot is, say, 0.4, we would expect him, just by randomness, to have some intervals when he makes a fairly large number of baskets in a row.

V. The question is whether the streaks that are seen for real basketball players are explainable by randomness alone.

A. One possible way to analyze the question of whether streaks are explainable by randomness alone is this: Suppose that when a player makes a basket, his probability of making a basket on the next shot is higher than his average.

1. Most real data of this sort do not indicate the reality of a hot hand.

2. As usual, our strategy is to compare the data that we find with distributions of data that we would expect to arise from randomness alone, that is, that would arise under the assumption of no hot hand.

3. If the data are so extreme that they and their more extreme versions would happen only rarely given the assumption of no hot hand, we would take the data as evidence that there is a hot hand phenomenon.

4. The statistical test used to measure rarity in this example is called a *chi-square test*.

B. In the example considered in the lecture, the data were not sufficiently extreme to reject the assumption of no hot hand. Thus, the data do not warrant the conclusion that there is a hot hand.

VI. Trying to distinguish randomness from some other cause is difficult.

A. There are many ways to look at a string of data and ask whether the string is explained best as a random process or as a process that is influenced by something internal to it.

B. There are many ways of looking for patterns.

C. There is room for interpretation.

VII. These two examples of sports analysis have brought up many statistical issues.

A. Trying to find the greatest hitter in baseball history brought up questions about the relationship between what we measure, what we want to know, and how to compare performances in different eras or under different circumstances.

B. The hot hand issue brought up fundamental questions about the nature, meaning, and measure of randomness.

Readings:

Jim Albert and Jay Bennett, *Curve Ball: Baseball, Statistics, and the Role of Chance in the Game*.

Stephen J. Gould, *Full House: The Spread of Excellence from Plato to Darwin*.

Michael Lewis, *Moneyball: The Art of Winning an Unfair Game*.

Questions to Consider:

1. What method would you use to select the greatest athlete of the 20th century? (I choose the 20th century rather than all time because we would not have much reliable data on earlier athletes.)
2. If randomness with a certain probability really accounts for sports performance, does that lessen the interest in watching sports? Would it change the approach to sports psychology and the training strategies?

Lecture Sixteen—Transcript
Sports—Who's Best of All Time?

Welcome back to *Meaning from Data: Statistics Made Clear*. Today we're going to take on the statistical challenge of analyzing sports statistics, which is a sport of its own. Statistics are, first of all, extremely important in sports, particularly in making the managerial decisions. I mean important managerial decisions, like the ones that you make when you're sitting around at home on your couch and a bunch of people are there, and you're watching some game, and you're helping the manager make a decision about what he or she should do next in the game. These are very important.

But not only are they important in that really useful work, but it's also true that these sports statistics are wonderful illustrations of statistical ideas that have applications well beyond sports. So, I think there's no question that looking at sports statistics issues is a wonderful way to both learn about statistics, and also have some fun.

So, in this lecture, we're going to pose a couple of sports questions. The first important question that we're going to pose is, "Who was the best hitter in the history of Major League baseball?" It sounds like an innocuous question; it sounds like a simple question. Because it sounds as though it's a question of just saying, let's look at the measure of hitting—namely, the batting average—and decide who has the best batting average, and that is the answer.

But, in fact, it brings up all sorts of basic principles and basic challenges about statistics. The first one is to clarify the question, "Who is the best hitter?" What does that question mean? What evidence are you going to accept as supporting the contention that a particular individual was the best hitter? So, the question forces us to clarify the relationship between what we measure and what quality we're trying to actually describe. That, you see, is extremely fundamental both in statistics and the world in general. Although, at first, we might think that the question really amounts to who has the best batting average—who has the highest batting average in history—we might think that is the question, but let's take that as a provisional issue and see what happens.

Well, first of all, right off the bat, we instantly come up to an issue about baseball averages. That is, certain players, such as the pitcher, don't really bat well. The batting average of a pitcher is inconsequential, and that shouldn't be related to anything. Of course, that pitcher is not going to win the batting championship anyway so maybe that doesn't matter. But the first question is, "What does the batting average actually measure?"

The batting average measures the fraction of the times that a batter comes up to the plate and the pitcher throws the balls to him and he tries to get a hit. If he successfully hits and goes to first base, then that's a successful hit.

There are lots of wrinkles to this. For example, if he makes a sacrifice fly, then that doesn't count against him. If he walks instead of gets a hit, then that doesn't count as an at-bat, and there are other wrinkles. But basically, the idea of a batting average—which, by the way, is always a decimal point and then a three-digit number, so .250 is 25%, and it's called .250 in the parlance—so, the point is that the batting average is telling us what fraction of the time the hitter actually gets a hit, basically speaking. But that is the first issue of what it is that the batting average actually measures.

The second question that you want to know in trying to say who has the best batting average is, "Do you mean the best batting average over the entire career of the player, or do you mean the best batting average over an individual season?" Suppose that the hitter went into his dotage and had a lot of low batting average seasons at the end of his career. Well, maybe he was really the best hitter, but it was earlier in his career. So, let's concentrate, then, on the question of a single season; so, we'll say who the best hitter is in a single season. That's just a decision we make.

The next question is, "Who are going to be contenders for this crown?" Suppose you have a batter, and it occasionally happens that a batter will only bat one time during the season, for some reason; maybe a pitch hitter comes in and just hits one time. If that person got a hit, then that person would have a perfect batting average and would throw off the whole scale. So, we have to make another decision that only players who have quite a few hits during a season should be contenders for this crown. We'll make the arbitrary decision to say that the person had to have at least 80 at-bats to count as a contender for the crown.

So, with these assumptions, let's just see how we do looking at the highest batting averages in the history of baseball, here on this chart.

The highest batting average occurred in the year 1901, and it was a batting average of .426. This is an amazingly high average, by the way. Here, we have a .423 in 1924; a .420 in 1922; and so on.

Look at the years in which these very high batting averages took place. Look at all these years. Can you see these years? Look at all of these years, 1901, 1924, 1922, 1911, 1912, 1911, 1920, 1941, 1925, 1923, and so on. Here's a .400 batting average of Ty Cobb in 1922. The batting averages you see are in order. Notice that all of these highest batting averages took place before the year 1942. The year 1941 was the last year in which any batter hit .400. It's true that down here at the very bottom, we have 1994 where Tony Gwynn hit a .394 average; but all of the previous ones above that were 1941 and before.

Well, when we see statistics like this, we have to ask ourselves what's going on. Because are we saying that all of the greatest hitters in baseball all played baseball before 1941? That's not a tenable conclusion. We seek to find an explanation for why it would be that all of these high batting averages occurred so early in the century.

Well, let's look at the batting averages over the years. One explanation that we might think of is, maybe in the early years, the batting averages in general, over all of the players, were much higher for some reason. The balls were different; the fields were different—something. It may be that all of the batters had higher batting averages and, therefore, it's not a fair contest.

So, what we can do is look at the data. These are the average batting averages for every year from 1880, up to the year 2004. You see in the year 2000, the batting average was about .266. Notice that this seems like sort of a complicated scatter plot, and it's all over the place. But, as a matter of fact, if we look at it a little differently, if we put it on the chart that shows where the vertical axis goes from a zero batting average to a perfect batting average, we can see that, actually, there isn't a great deal of difference between the years.

If we look carefully at our chart, we'll see that, in fact, in the year 1920, the batting average was within 2 points of the batting average of the year 2000. So, yes, there are differences—particularly this year was a very high year, a very early year before 1900, and even these years and the late '20s had very high years—so the average batting averages were higher, so we might expect some people to do extremely well.

In fact, though, there is an anomaly here. If we look at the years 1920 versus the year 2000, both of those years had almost the same average batting average over the whole league; but look what happens. If we look at the top batting averages for 1920—George Sisler, Tris Speaker, Joe Jackson, Babe Ruth—we can see that all of these people have higher batting averages than the very highest batting average in the year 2000, even though the batting average overall was the same. Now, this is a little bit odd. Why would there be higher batting averages in 1920, versus the year 2000? It doesn't seem that our guess of higher batting averages overall are accounting for that difference.

So, we have to think more. Here's what we do, as we've seen before in several lectures. That is, we notice that in the year 2000, the batting averages are more closely clumped to the average than they were in the year 1920. In the year 1920, they were more spread out. The standard deviation in 1920 was greater than the standard deviation of the batting averages in the year 2000.

That gives us an idea about how we can compare batters from eras where there are two different standard deviations. Instead of just talking about their absolute batting average, we could talk about how many standard deviations above the mean the batters were from one year to the next. In fact, by doing that, we could give everybody a measure. Instead of saying that Joe Jackson has a batting average of .382, we could say, "How many standard deviations above the mean of his year—1920—was his batting average?" In his case, it was 2.36 standard deviations above the mean.

Moises Alou had a batting average of .355. You see, that's a lower number; .355 is lower than .382. Yet, because the batting averages overall were more compressed, his batting average was also 2.3 standard deviations above the mean. So, comparing batting averages on the basis of their relative standing, relative to their contemporary players, seems like a good way to measure quality of a player.

Now, look what's happened over the century. These data here represent the standard deviations of the batting averages for every year, from way back here, up to the year 2000, and beyond. The standard deviations are clearly declining over the century. You see that? These are for batters who batted at least 80 times. You can see that the standard deviation in the year 2000 was much less.

We might ask why the standard deviations are lower in later years. Stephen Jay Gould argues that batters—and, in fact, pitchers and fielders as well—have all moved closer to absolute human limits of perfection. As professional players use techniques, and get closer to that limit, their variation in performance must decline; therefore, the standard deviations decline.

Well, what that means is it gives us an opportunity to evaluate the quality of a batter relative to his contemporaries across the era, across the span of time. In that measure, for every batter, we can say "How many standard deviations above the mean was that batter?" We can take all of the batters in history and look at their standard deviations above the mean for that particular year, for every year, and here's what we get.

In this measure of standard deviations above the mean batting average for that year, we see that the winner would be Tip O'Neill, who, incidentally, I thought was a politician. In 1887, Tip O'Neill had a batting average of .435, which by the way, is amazing for anybody. The number of standard deviations above the mean was 3.58. So that was his z-score.

You may recall, with standard deviations, when you have a normal curve, the number of standard deviations away from the mean tells you how much of the data lie that close to the mean. Remember that 68% of the data lie within 1 standard deviation of the mean; 95% of the data lie within 2 standard deviations of the mean; and 99.7% of the data lie within 3 standard deviations of the mean.

Here we have an example of a player who was 3.58 standard deviations above the mean. That corresponds to a fraction of 1 in about 5,000 that would be that far away from the mean in a normal distribution, which batting averages are, roughly speaking.

What's interesting about this way of comparing the batting prowess of people over time is, look what you have. The number one player is Tip O'Neill from 1887, whereas the number two player is Tony Gwynn of 1994, who had a batting average of .394, which is an amazingly high batting average for a modern era player and a z-score of 3.56. Then we move on to George Brett in 1980, and then back to 1884 for Fred Dunlap. Then we get to Ted Williams, Rod Carew,

Tris Speaker, and so on. In this evaluation, we get a different list of people from the people we got when we were just looking at their high batting averages.

Here are the next 10, which include familiar names: Ty Cobb, Ted Williams again. We have a very modern player, Ichiro Suzuki, 2004, who had a very high number of standard deviations above the mean.

But we must ask ourselves the basic question, "Are we measuring the qualities that we really want to describe as the best hitter?" You see, batting average is one quality, but when you get a hit, there are several different ways you can get a hit. You can get a single—but you could get a double, a triple, or a homerun. You should get extra value for hitting a double, or a triple, or a homerun than just getting a single.

Well, we could devise a statistic to measure that quality and, in fact, this is done. What we could do is give twice as much value for a double; three times as much for a triple; and four times as much for a homerun. This number is called the *slugging percentage*, even though it's not a percentage. It's not a percentage because it's out of 4 instead of out of 1. With the slugging percentage, if you got a homerun every single time you came to bat, you'd get 4 points each time, so you'd get a total of 4; that would be perfect. But people who often hit multiple-base hits will have a higher slugging percentage.

This is a different way of deciding who the best hitter is, by looking at slugging percentages. We can have an interesting analysis by looking at that evaluation of hitting prowess. Listing people in order of their slugging percentage, we get a different picture.

Here, we have number one, Barry Bonds, with a slugging percentage of .863; Babe Ruth, in 1920, had a slugging percentage of .847. These are season-by-season, so the same player appears different years because we're looking at the slugging percentage for a season. Here we have Barry Bonds; Babe Ruth; Babe Ruth; Barry Bonds; Barry Bonds; Babe Ruth; Lou Gehrig; Babe Ruth again; Rogers Hornsby; Mark McGwire—a relatively modern player; Jeff Bagwell; Barry Bonds again; Jimmie Foxx; Mark McGwire; Babe Ruth; Babe Ruth; Sammy Sosa; Ted Williams; Babe Ruth again. So, Babe Ruth occurs an amazing number of times on this list of slugging percentages.

One wrinkle to this list, which I think is rather interesting, is Barry Bonds. Look at Barry Bonds. He has numbers here in the year 2001, 2004, 2002, and down here, 2003, all the years—2001 through 2004—he is in this list of the top 20 sluggers.

What's really interesting about that fact is that up until Barry Bonds was 35 years old—he played professional baseball starting at age 21—from years 21 to 35, he never had a slugging percentage that reached the 700 level—not the 800 level, but the 700 level. He never reached the 700 level. But in the year 2001, he got much better and had the all-time record for slugging percentage and, in fact, in 4 years. This is a remarkable statistical accomplishment, to have had a certain level of performance from age 21 to 35, and then to have a big leap, particularly late in the career of a baseball player.

When we look at statistical data that presents an apparent anomaly, then what we tend to do is to look for a lurking variable—that is to say, something that's going on that's not apparent in the data that's being presented.

Looking at baseball statistics, we have another kind of issue associated with who is the best hitter. That is, if you are talking about contributions to the team, just getting hits is not the only contribution for the offense. In fact, it's very important to get a walk. If you can get on base, whether it's by a hit or by a walk, you're on base and in a position to help with the scoring. So, in fact, measurements such as on-base percentage are very important, and managers will look at issues, such as combinations of on-base percentage, and slugging percentage, and batting average, and make formulas that try to assess the real quality to the contribution of the offense of a player based on these kinds of statistical evaluations.

So, the theme of this first exercise was, one of the important ideas, was to understand the correspondence between what we measure and what it is we're actually trying to ascertain. This is a very important issue in statistics.

We're now going to turn to a second sporting issue. This is an issue that all of us who have actually played sports have experienced personally, and that is the idea of having a streak. A streak occurs when you're in "the zone"; you've got that "hot hand"; everything you do is right; it just works, and you just feel great.

People have been trying to study the question of the hot hand, where a player's on a streak and just can't miss. The question that we want to ask is, "Is there really such a thing as a hot hand?" In other words, there are streaks, yes, there are definitely streaks. If you have, for example, a basketball player, there are streaks where that basketball player will hit many shots in a row.

But the question is, "Are those streaks better described by saying that that player is in the zone, and has a hot hand, and is actually doing much better during that time? Or, alternatively, is it like flipping a coin?" When you flip a coin, it lands heads or tails; you flip it again, it lands heads or tails. It may be that you flip it 10 times and 10 times in a row, it comes up heads. You wouldn't want to say, "Wow, that coin really has a hot hand now." You would say, in fact, "It's a result of the fact that if you do things randomly, you are going to sometimes get a long sequence of results that appear not to be random." You see?

So the question is, "How are we going to tease apart these two possible descriptions of a phenomenon that is real?" Namely that, yes, indeed, this basketball player does sometimes have many baskets in a row. Do we describe that better by saying it's just explained by randomness, that the player has a certain random possibility of making a basket or not each time, but just because of randomness alone, it turns out that this streak happened?

This is a very deep question, the question of distinguishing randomness from something that involves intent. So, let's try to think about how we would go about comparing the random to the things that are caused for a reason.

In order to do this, let me ask you to look at the following list of H's and T's. I know this is a long list of H's and T's, but I want you to look carefully at this list. I'm going to tell you that both of these have the same number of H's and T's, but they were created in two different ways. One of these strings of H's and T's was created by taking an actual coin, flipping the coin, and then just recording H and T according to how it came up. That's the random list.

The other list was created by a person who was just asked to write down a random list of H's and T's that would come up randomly. It's a simulation of randomness that a person did, but didn't actually flip the coin. This is a wonderful question to ask: Can you tell the difference between these two lists? In a way, it seems impossible, because the person intended to try to make it random, and they sure look random. Don't they look random here? These all look random; these look random.

But, as a matter of fact, looking at those lists, you can be very, very confident that the second list here is not created by a random process. The reason that you can be so confident that this second list is not from a random process is that it does not have streaks. Streaks are rare events, but they're absolutely expected if you have a long sequence. You see what I mean? You have to expect the unexpected; you have to expect the rare. That's one of the properties of randomness—that a certain number of times, you will get a long streak of six or seven in a row.

Look up here. This is created by flipping coins, and look at this. Here you have six H's in a row. Often, a person who's intentionally writing down things that he or she is told to appear random would think, "Oh, if I put six H's in a row, that doesn't seem very random to me." You see? You can actually tell the difference between random and trying to be random by looking at that kind of distinction.

Well, what this is saying is that you expect streaks to occur by random chance alone. Let's look at how we would go about trying to analyze streaks in basketball. Here we have a series of 63 hits and misses from a player in a basketball game. The question is, "Can we try to evaluate, from looking at that streak, whether or not it's the result of randomness, or whether it's an indication that there are streaks?" How would you go about distinguishing the two?

Here's an idea. Suppose that a person who made a shot was more apt to make another shot because, you see, they were in the zone. That's what a streak really means, that somehow making one helps you to be better at making another. So, what we could do is look at how often the person successfully made the basket after making a previous basket, compared to how many times the person made a basket after a previous miss. You see?

So, hit means that it was a good basket; M means it was a miss. What we can do is look at this. After a hit, how many times was it a hit? 16. After a hit, how many times was it a miss? 12. After a hit, there were a total of 28 occurrences of which 16 times it was a hit, and 12 times it was a miss. After a miss, on the other hand, there were 13 occurrences where this player got a hit, and 21 times when it was a miss.

So, this gives us the same kind of a square of data that we saw earlier in the lecture about the law when we were talking about the discrimination case. The question here is, "Is your probability of getting a hit after a hit different from your probability of getting a hit after a miss, or is it about the same?"

When we actually do the analysis, we can see that there's always going to be some probability involved in saying how often you're going to get hits and misses; there will be some distribution. If it's just random, you would expect the same proportions of hits and misses after a hit as the proportion of hits and misses after a miss. But, you'd also expect a certain amount of variety just randomly.

Once again, the chi-square statistic is the method by which we compute how rare the disparity after hits and misses is in this data, and we see that the p value for this is .14. Generally, the p value, in order to be persuasive as a measure that something happened not by chance alone, we would require that the p value—which is the probability of something that rare or rarer occurring—to be less than 5%. So, since this is 14%, we would not reject the null hypothesis that, in fact, having a hit or a miss does not affect the frequency with which you then get a hit or a miss on the next opportunity.

These issues that we've looked at have brought up a lot of basic statistical questions. One is, asking the right questions and using the appropriate data to evaluate and support evidence for answering a particular question that you're interested in. The second question about the hot hand brought up the very fundamental question about nature—namely, about the meaning and the measure of randomness. I'll look forward to seeing you next time.

Lecture Seventeen
Risk—War and Insurance

Scope:

In World War II, the serial numbers on captured Mark V German tanks were used to deduce the number of Mark V tanks produced altogether. We will use that scenario to introduce a variety of methods of inference and to analyze how different plausible methods might be compared for expected accuracy. Risk closer to home occurs when we deal with insurance. Insurance is an industry based on probability. In determining whether buying an insurance premium or extended warranty on a product is a good investment, we deal with the statistical ideas of expected value and the distribution models of product lifetimes.

Outline

I. In World War II, one of the challenges of the Allied intelligence officers was to estimate the strength of the German fighting machine.

- **A.** In particular, one wanted to estimate the number of tanks that the Germans had manufactured and were using in battle.
- **B.** During World War II, when a German tank was captured, analysts noticed that the tanks had serial numbers and it appeared that the serial numbers were consecutive, starting with 1 and increasing as each new tank was built.

II. Statisticians approached the situation as a statistics question. We know some information about part of the population, and we wish to infer information about the whole population.

- **A.** We assume there are a certain number of tanks in the German army, numbered from 1 to N.
- **B.** We assume that the tanks captured are a random sample from the whole population of tanks.
- **C.** We would like to estimate the total number of tanks.
- **D.** We'll look at possible *estimators*, that is, methods or strategies for calculating an estimate.

III. Let's do a specific example. Suppose we've captured tanks whose numbers are {68, 35, 38, 107, 52}. What estimate for the number of total tanks would we make?

A. One idea might be to take the mean of the five numbers, double the result, and subtract 1—giving an estimate of 119.

B. Another method for estimating the midpoint of the numbers 1 to N would be to take the median value of the sample. Then, we would double that, giving an estimate of 104. This estimate is clearly too small, because it is less than the number on a tank we actually captured.

C. In fact, both of these strategies can produce an estimate that is less than the highest numbered tank we have actually captured.

D. Another strategy for guessing the midpoint of the total tank numbers would be to add the biggest and smallest numbers that we've captured (107 + 35) and take their average (71).

1. We would double that average to get our estimate (142).

2. This method always produces a number bigger than the largest in our sample. Thus, this estimator doesn't suffer the previous method's flaw.

E. These various estimators each have some intuition behind them, but there are actually other strategies that are superior.

IV. We'll discuss what qualities different estimators can have in order to guide our decision about which method, among reasonable-sounding methods, to use.

A. One feature that a particular method of generating estimates (an estimator) can have is that the method maximizes the probability of choosing the sample we actually got.

B. $N = 107$, the maximum number on a captured tank, would maximize the chance of capturing our collection. However, our intuition tells us that a good estimator would estimate a larger number of tanks than the largest number we have actually captured.

C. In this case, the *maximum likelihood estimator* (as such an estimator is called) doesn't seem to be reasonable.

V. One property that we might want a method to have is that in performing the method many times and taking the average of the estimates the method produced, we would, on average, get the true number of tanks.

A. This average of the estimates is called the *expected value* of the estimator.

B. An estimator whose expected value is the correct value is called an *unbiased estimator*.

C. For example, the 2 × sample mean – 1 estimator is unbiased because the sample mean is an unbiased estimator of the population mean. But we saw that this estimator had other drawbacks (namely, producing values that are definitely wrong).

D. Can we think of a strategy (an estimator) that is unbiased and doesn't give us answers that are definitely wrong? Yes, we can.

1. Estimate $= \dfrac{k+1}{k}\max(x_1, x_2, x_3, \ldots, x_k) - 1$. In our example, $\dfrac{5+1}{5}\max(68,35,38,107,52) - 1 = \dfrac{6}{5}107 - 1 = 127.4$.
2. This method is unbiased. That is, if we use this computation on every possible sample, then the average of those estimates will be N.
3. The expected value of the estimator is N, the quantity we are trying to estimate. That is, we can't be sure that the estimator will give the correct value, but on average, it will give the correct value.

VI. Another desirable quality of an estimator is to have as small a variance as possible, because that would mean that the estimator is, on average, close to the true value.

A. It turns out that the estimator previously defined (that multiplies the maximum of the sample by $\dfrac{k+1}{k}$ and subtracts 1) is the *minimum variance unbiased estimator*.

B. This study of tanks brought up the idea of expected value, which is a central idea in the risky business of buying and selling insurance.

VII. One of the practical ways of tempering the vagaries of risky life is through insurance.

A. The whole concept of insurance is based on statistics.

B. We can view insurance as a game of chance.

C. Our decisions on whether to buy an extended warranty, health insurance, or other insurance are best based on understanding the distributions of the foreseen calamities that the insurance is aimed to mitigate.

D. But most people are generally not good at gauging large numbers or rare events.

VIII. We can view an insurance company that sells insurance to many people as playing the same game with many people.

A. The company needs to consider the distribution of possible payouts.

B. To illustrate, we'll consider the following game: We shuffle a deck of 52 cards. You draw a card. If it is the queen of spades, the insurance company pays you $100.

C. If the game is played by 1,000 customers, the distribution of the number of payouts is binomial, with $p = 1/52$ and $n = 1{,}000$. Thus, we would expect about 20 people out of 1,000 to be winners.

D. For that distribution, 98.7% of the time, the number of winners (payouts) will be between 0 and 29.

E. If the company has enough money on hand to settle 29 claims, it will be 98.7% sure that it won't run out of money.

IX. Now, let's take the example of extended warranties on electronic items.

A. The distribution of the time to failure on a new electronic item is not just a smooth distribution that declines over time. Instead, it may be more like a bimodal distribution.

1. Electronic items that are going to fail sooner than average will tend to fail almost immediately because they were never made properly.
2. But if they are working after a few months, then they are likely to continue to work until much later, when they come to the end of their expected life.

B. Thus, the distribution is bimodal, with one peak near the beginning of use and another after some prescribed length of service.
1. The extended warranty really only covers the period between the end of the manufacturer's warranty and the time when the manufacturer thinks that second peak will occur.
2. That tends to indicate that those kinds of insurance policies may be particularly poor values.

X. From risks in war to risks in insurance, statistical analyses pay good dividends.

Readings:

David S. Moore and George P. McCabe, *Introduction to the Practice of Statistics*, 5th ed.

Questions to Consider:

1. Suppose you captured tanks numbered 25, 64, 253, 135, and 85. Assuming that you were selecting tanks randomly from ones numbered sequentially, what would be your best guess for the number of tanks that exist altogether?
2. None of the methods we talked about concerning guessing the number of tanks made use of the order of the numbers of the captured tanks. Under the conditions of the question, that is, that the tanks are captured randomly, could the order in which the tanks were captured be a relevant factor?

Lecture Seventeen—Transcript
Risk—War and Insurance

Today's lecture concerns two risky businesses: war and insurance. We'll start off in World War II. In World War II, one of the challenges of the Allied Intelligence officers was to estimate the strength of the German army. In particular, one of the challenges was to estimate the number of Mark V tanks that the Germans had manufactured and were using in battle. Statistics played a heroic role in this quest for the number of tanks. Our lecture begins on the battlefields of World War II.

During World War II, when a German tank was captured, people began to notice that the tanks had serial numbers on them. After capturing several of the tanks, the Allied Intelligence officers began to suspect that the serial numbers were simply consecutive numbers starting at 1, and increasing as each new tank was built. Possibly the German order was evident in their choice of serial numbers on the tanks.

So, statisticians approached this situation as a statistics question. Just think about it. This is exactly the realm of statistical inference, namely, we know some information about part of the population, and we wish to infer information about the whole population. We know that there are tanks numbered from 1 to N. Let's call N the total number of tanks in the German army, these Mark V tanks.

Let's make the assumption that the tanks are captured as a random sample from the whole population of tanks. What we would like to do is take that information about that random sample and estimate the total number, N, of tanks in the whole German army.

So, what we're going to do in this lecture is look at different methods, or strategies, for calculating that estimate. A method for estimating something is called an *estimator*. So, we're going to look at different estimators for this problem.

So, let's start off with a specific example. Suppose we've captured five tanks, and the serial numbers on those tanks were 68, 35, 38, 107, and 52. The question is, "What estimate should we make for the number N of the total number of tanks in the army if that was the sample we had?"

Let's think about it as an abstract problem. Namely, the problem is exactly the same as if we had a big bag of poker chips, and each chip had a number on it—1, 2, 3, 4, 5, up to some number—but we don't know how many. Then we reached into the bag and took out five of them, and then the numbers on those chips were those five numbers that we just said. The question is, "How would we make an estimate for the total number of chips in the bag?" It's exactly the same question.

Let's think about different strategies of making an estimate. What would be a reasonable way to do it? One idea that we've seen many times in the past lectures is that when we have a sample from a population, if we take the mean of the sample, then the mean of the sample is an approximation to the mean of the whole population. So, maybe we could use that concept to make an appropriate guess for the number of tanks.

We discussed this kind of idea many times before. For example, when we estimated the average height of males by taking a sample of 5, or 10, or 100 males and taking the mean of that sample, that was an estimate of how tall an average male was in the country. The same thing applies to the proportion of voters for a different candidate. We'd take a sample, see what the sample was like in terms of proportion of voters, and then we estimated what the total was like on that basis. The idea is that the mean of the whole population is estimated by the mean of the sample.

So, since the mean of the numbers from 1 to N—taking all the numbers 1, 2, 3, 4, 5, up to N—what's the mean of that number? It's N + 1 divided by 2. Doubling the sample mean would give 1 more than N. Let's work it out here. The sample mean, x-bar—the notation for when we take the mean, add up the numbers, and divide by the number we have—will be approximately N + 1 divided by 2, which is the whole population mean. Just doing the algebra, we see that N—which is the total number of tanks we're trying to estimate—is going to be approximately equal to 2 times the sample mean minus 1.

So, in our particular example, let's go ahead and do the arithmetic. This 2 times the sample mean minus 1 is 2 times this quantity right here, which is the sample mean, adding up the numbers and dividing by 5, subtracting 1, gives us the estimate of 119. This doubling the sample mean minus 1 is an estimator. It's a strategy for estimating the number of tanks. It makes sense because it reflects the fact that the sample mean should be close to the actual mean.

But there are other methods we might consider. Let's try one. Another method to estimate the midpoint of the numbers from 1 to N would be to take the median value of the sample. Once again, the median value of the sample should give us a number that is somewhat near the middle of the total population numbers, because we're taking this random sample. So, let's go ahead and try that.

So, doing the median number, we could just take the median number, and basically multiply it by 2. That would be another estimator. For our particular example, the median—which is where you put the numbers in order, and take the middle value—of the numbers 35, 38, 52, 68, 107, that is the middle value, is 52, and 2 × 52 is 104. So, the estimator of taking the median value and multiplying by 2 would give an estimate of 104 tanks in the German army.

Hmmm. Let's think about this. That is a ridiculous estimate. That's clearly wrong. Why is 104 clearly wrong? It has a basic flaw. The basic flaw is that it estimated a number—namely 104—that was lower than the highest serial number of a tank that we actually captured. So we know that this strategy, the double-the-median strategy, doesn't necessarily give us an estimate that is credible. You see? In our actual example, we saw that it gave us a number 104, and we know there are at least 107 tanks.

Let's consider whether or not our previous method suffered from that same flaw. In the previous method, remember, we took the mean of the sample, doubled the mean, and subtracted 1, to make our estimate. Would it suffer from that same possibility of producing a number that was actually less than the biggest tank number that we actually captured?

Let's think about that. Suppose that we captured an additional tank, and the additional tank's number was 300. So, now we've captured six tanks, and here are their serial numbers. Then, if we apply the estimator method of doubling the mean and subtracting 1, look what happens. We get 2 times the mean of all these numbers—which is adding them up and dividing by 6—minus 1, which is 199. That's clearly wrong—because, once again, 199 is less than the biggest tank number that we actually captured.

So, there's a flaw to our method. Even though the logic seemed very good, there's a flaw—namely that either of these two methods have the property that it's possible for our estimator to yield a result that's actually less than the largest number we've actually captured, and, consequently, is definitely too small.

So, these two estimators, although they're conceptually good ideas, aren't the best possible way to do it. Here's a different strategy. We could guess the midpoint of the total number of tanks by taking the biggest tank number we captured and the smallest number we've captured, taking their average, and doubling that. You see? So, this would be doubling the midpoint between the highest and the lowest, and that method at least always produces a number that's as large or larger than the biggest number in our sample; so that's good. At least it doesn't suffer from that flaw of the previous methods, which is to give an estimate that is definitely too small.

These various estimators each have good intuition behind them, but, actually, there are other strategies that are even better. What I'd like to talk about now are the qualities we would look for among different methods, among different estimators, that might guide our decision about which reasonable-sounding strategy would be the best one to use.

In a way, there's no single right answer as to what method will generate the best estimate, but we can ask the question, "How would we evaluate strategies? How could we make statements about various properties that a method, an estimator, has that would make it more or less desirable?"

So, let's use the tank example to ground a discussion of desirable qualities of methods of statistical inference. One feature that a method could have for generating estimates—that is, an estimator—is that it could maximize the probability of choosing the sample that we actually got. Let's follow it through here.

For example, suppose there were 500 tanks in the German army. It would be possible—but very unlikely—to have randomly captured the five tanks of our original examples—numbers 68, 35, 38, 107, and 52. The reason it's so unlikely is because all those five numbers are so much smaller than the 500 possible numbers; therefore, it's unlikely that you would have found only those relatively small numbers. In fact, if there were 1,000 tanks, getting those particular five numbers would be even less probable. But if you have fewer tanks then the probability of getting those five numbers becomes larger.

So, we might ask ourselves, "What number of tanks in the German army would maximize the probability of our having gotten those exact five numbers, the numbers we actually got?" For example, if there were 200 tanks, we'd compute the probability of getting those five numbers. If there were 150 tanks, we'd compute the probability of getting those five numbers. In theory, we could do that analysis for every possible number of tanks, and see which number gave us the largest probability.

I'll tell you the answer because the answer actually is very simple, but very unsatisfying. The answer is that if the number of tanks is 107—namely, the largest number among the tanks we actually captured—then the probability of choosing those particular five numbers is higher than any other number of tanks you could have.

The reason is very simple. If you're just selecting any particular set of five numbers from a total set of numbers, the smaller number of numbers you start with, the better chance you have of picking any set of five. But our intuition tells us that we're not likely to have captured the last-built tank so our intuition is that a good estimator would somehow have to estimate a number that's larger than the biggest number among the tanks we actually captured.

So, in this case, this *maximum likelihood estimator* (as the estimator is called)—that is the one that maximizes the probability of having gotten the sample we actually got—doesn't seem to be reasonable because it would just guess the biggest number that we got.

So, statisticians have defined several qualities that an estimator may or may not have as a way of formalizing our intuition about what makes a good estimator. One property that we might want a method to have is that if the method is performed many, many times—that is, if we took many samples of five tanks, and in each case, evaluated our estimate of the total number of tanks, using whatever method we're discussing—on average, the estimates that it produced would, in fact, be the true average of the number of tanks.

The average—by which I mean actually the mean of the estimates—is called the *expected value* of the estimator. The expected value of the estimator is the mean of all possible answers you would get by using that estimator for all possible samples in this case of five numbers.

An estimator whose expected value is the correct value is called an *unbiased estimator*. If we use the estimator that just gives us the maximum of the samples that we chose—in other words, if we use the estimator where we just capture tanks, and we guess that the number of tanks is the biggest number in that group—then that estimate is certainly not an unbiased estimator.

Let's be concrete about this, and suppose that we have a certain number of tanks in the army, N, and we randomly choose five tanks, and we compute the estimator. For example, our estimator is to take the maximum of those five numbers.

If we imagine doing this for all samples of size five from the total N tanks in the whole army, then if we take the average, the mean, of all those maximum values, we certainly wouldn't get N. Because no estimate, using that method, would ever be bigger than N, and most sets of five would have the maximum number smaller than N. So certainly, the mean of all those sets of five, of the maximums of the sets of five, would certainly be less than that. So, that estimator, just taking the maximum of the sample, is not an unbiased estimator.

On the other hand, the estimator that uses 2 times the sample mean minus 1 is an unbiased estimator. It had other defects, but it is an unbiased estimator because, as we saw many times before, the sample mean is an unbiased estimator of the population mean, meaning that when you take all the samples of size five, the mean of all those means is the same mean as the whole population.

But, we saw that that estimator—2 times the mean minus 1—had other drawbacks. Namely, it produced values that were definitely wrong, not big enough; but on average, at least it's right. On average, it's right, but sometimes it's definitely wrong.

Can we think of a strategy, an estimator, that's unbiased and yet never gives us an answer that's definitely wrong? The answer is, yes, we can. Here is another estimator that will turn out to be an unbiased estimator, a very good estimator. Here is the idea. The idea is that when we choose a certain number of tanks from our randomly selected group, we would expect the highest number of the tank that we actually captured to be somewhat less than the maximum number. What we're going to do is try to estimate how much less.

Let's look at an example. Suppose that the actual total number of tanks in the population were 1 through 7, and we selected 3 at random. If the 3 at random were evenly spaced, our maximum number would be 6. So, if we take 6 and multiply it by some fraction to get it up to 7—in other words, we're trying to make it bigger—we're going to guess a number bigger than 6, and we're asking ourselves, "How much bigger should we choose?" The answer is, if we multiply 6 by 4/3—because you basically are dividing this range of 1 to 7 into 4 parts—you're going to take 1/3 of them and add an extra one, then subtract 1, you get 7.

That's a little bit of vague intuition about the reason for choosing the estimator, of taking the maximum tank number that we actually capture and multiplying it by k + 1 divided by k, where k is the number of tanks in our sample that we captured.

So, we're trying to estimate how much to augment the maximum tank we captured in order to make a good estimate of the total number of tanks in the army. So, we'll multiply it by a number that's slightly bigger than 1—namely k + 1 divided by k—and multiply it by the serial number on the biggest tank we captured—that's the maximum—and then subtract 1 because that fits in with the previous example and gives us a good estimate.

So, in our example that we had of these five captured tanks, we would multiply that maximum number—namely, 107—by 6/5 because we captured five tanks—so k is 5—so we have 6/5, times that maximum 107, minus 1, gives 127.4. So, that would be our estimate for the number of tanks using this estimator.

This method of estimating tanks is an unbiased estimator. That is, if we use this computation on every possible sample, then the average of those estimates will, in fact, be N. The expected value of the estimator is N, which is the quality that we're looking for in an estimator for it to be an unbiased estimator. Now, of course, we certainly can't be sure that it's going to give the correct value, but the mean of the answers that it gives will, in fact, be the correct value.

We've done some simulations here. Here was a case where we took 3,000 tanks as the actual number of tanks, and from those, we took samples of size 100, meaning that we imagined that we captured 100 tanks at random. For each of those captured tanks, we looked at the maximum number of all those 100, multiplied it by 101/100, to get a bigger number, and then subtract 1 to get an estimate. We did that whole process 1,000 times, just as a simulation, a computer simulation. We did it as though we were fighting World War II 1,000 times.

Here were the estimates that we got, given as a histogram. The minimum was 2,803 of our estimate, when the reality was 3,000. The median was 3,010. The maximum was 3,030. The mean, as prophesied, was extremely close, at 3,001. Because, remember, it's an unbiased estimator, meaning that on average, the mean of the estimates that it gives, tend to be exactly right, and, sure enough, here is an example of where it was. This is a simulation that suggests that it is, in fact, an unbiased estimator.

Another desirable quality we want in an estimator is to have as small a variance as possible because that would mean that not only are the estimates close to the correct answer, but they don't vary too much from each other. So, we actually are looking for an estimator that is not only an unbiased estimator—meaning that on average, it gives us the right answer—but also that it has small variance, meaning a small standard deviation. This estimator of taking k + 1 divided by k, times the maximum, minus 1, is an unbiased estimator that has actually the smallest variance of estimators we can have.

In our simulation here, we had a standard deviation of 30, which is saying how spread out our guesses actually were. The idea being, of course, that if the variance is small, then we have a high likelihood that when we use the estimator, we'll actually get an estimate that's close to the actual value of N. As I said, the estimator that we just talked about turns out to be the estimator that is the *minimum variance unbiased estimator*.

The study of tanks has brought up the idea of expected value, which is a central idea in the other risky business of buying and selling insurance. So, insurance is a way of tempering the vagaries of a risky life. The whole concept of insurance is based on statistics. We can view insurance as a game of chance, where we determine the value of insurance as the expected value of the return. But, of course, our decision to buy insurance—or an extended warranty on a product, or health insurance—is based on a lot of other issues besides just the expected value of the return. It's a method by which we can make sure that our risk doesn't include great calamities.

The problem with insurance is insurance always involves large numbers and scarce events, which people are not good at gauging in an evenhanded way. It makes it difficult for us to evaluate issues about insurance. For example, how much is an appropriate amount for an insurance policy to charge?

Let's look at an insurance company that is insuring us against choosing the Queen of Spades from a deck of cards. Suppose that we have a deck of cards, and the insurance policy is the following: People are going to take this deck of cards and choose a card. If they choose the Queen of Spades, then the insurance company has to pay them $100. This is very simple insurance.

The question is, "How much money should the insurance company have on hand to know that it can pay off all the possible claims? Also, how much should it charge for the insurance in order to make money as an insurance company?"

Well, our strategy is that we look at the distribution of possible numbers of winners out of these 1,000 players. It's possible that no one will choose the Queen of Spades. But it's possible that, in fact, all 1,000 people will choose the Queen of Spades, in which case the insurance company goes broke.

But, the point is that we can look at the distribution of the likelihoods of 1,000 people, of how many out of those 1,000 are likely to choose the Queen of Spades. It's going to be centered at about 1/52 of 1,000 because there are 52 cards in the deck, and each player of the 1,000 has a 1/52 chance of picking it.

But, we can see this distribution that shows us that, generally speaking, the number of people likely to choose the Queen of Spades will vary between about 10 and 30. So, if we want to be 98.7% sure of staying in business, we can make our insurance company have enough money on hand and charge high enough premiums so that it can pay the claims for 29 or fewer winners. Because 29 or fewer winners—winners being people that you have to pay off—will account for 98.7% of the time.

When buying insurance for products—for example, extended warranties, when you go into a company and you buy an electronics item, they'll always try to sell you an extended warranty—it's important not only to know the average time to failure of your electronic device, but, in fact, the distribution of those times to failure.

Let me just show you a chart. Suppose that this is the chart of the time to failure of some electronic device that you're buying. You're trying to decide, should I buy the extended warranty? The advertisement will say, on average, the time to failure is right here, so you should buy the extended warranty. But, in fact, the distribution of times to failure may look more like this graph, in which case, this is the likely location of the two important moments. One important moment is the end of the manufacturer's warranty, which is when things fail because of some manufacturing defect. Then, once an electronic item is going, it works pretty well for a while, and then right about here will be the end of the extended warranty period, at which time a lot of these devices will break. At least this is the common experience of many people in trying to decide whether or not to buy the extended warranty.

It's been my pleasure today to talk to you about risk in the two areas of war and insurance.

Lecture Eighteen
Real Estate—Accounting for Value

Scope:

Tax authorities often need to set valuations for each house in the tax district. Because some of the houses have sold during the year, their market values are known; however, most houses were not sold. The challenge is to use the data about the sold houses to assess the values of all the houses. This situation is a classic example of statistical inference. Using multiple linear regression, we can find a formula that predicts the selling price of a house based on measurable quantities, such as square footage, number of bathrooms, distance from the city center, and so on. A case study illustrates how such multiple linear regressions are done.

Outline

I. The goal of this lecture is to give some sense of the types of issues that we confront when actually doing a real-life problem. Probably you will not feel the need to follow every detail, but you will get a sense of the rhythm of a multiple regression analysis.

- **A.** In this lecture, we confront the real-life problem of producing assessments of the market values of houses in a city.
- **B.** In most of our lectures so far, we have dealt with one or, at most, two varying quantities.
- **C.** Our goals here are to organize, describe, and summarize data when multiple variables are involved.
- **D.** We wish to know which quantities affect, explain, or are related to which others (and to what degree). Square footage, lot size, number of bedrooms, number of bathrooms, distance from city center, and other factors all influence the market value of a house.

II. The square footage seems likely to be the most influential single variable.

A. Our data consist of the square footages and the selling prices of a collection of 113 houses sold during the last year in our example city.

B. We first describe a *linear regression* using square footage as the *explanatory variable* and selling price as the *response variable*.

C. One step is to look at the two variables independently.

1. Specifically, we can draw a histogram and compute the mean and the standard deviation of the square footages and perform similar calculations for the sales prices.
2. The resulting graphs look similar and both have right skew.

III. We can visualize the relationship between these two variables by making a scatter plot of the two variables.

A. The cloud of points looks roughly linear going up to the right.

B. We can approximate the scatter plot by the least squares regression line.

1. The difference between the regression line's second coordinate and the data pair's second coordinate is the *residual*.
2. Squaring each such difference and adding them up gives a sum of squares. That is, we are taking the sum of the squares of the residuals.

C. The *least squares regression line* is the line for which the sum of the squares of the residuals is least.

1. Software can compute a formula for this line.
2. In our example, the equation is:
   Price = \$161 × square footage – \$63,600.

D. The slope of the regression line tells us how much change in the y variable is expected from each unit change in the x variable.

1. In other words, the slope tells us how much more we would pay with each additional square foot.

2. In our example, that slope is $161 per square foot.

IV. The view of how we are thinking of the data is summarized by: Data = Model + Residuals.

V. How well does that summary capture the actual data set?

A. We know that the values of the second coordinate (in our example, the house prices) vary.

1. The variance (the square of the standard deviation) of the house prices is a measure of how spread out those prices are.
2. Recall that correlation measures how closely two quantities move together. In this example, the correlation between square footage and house price is .835.

B. Now we see how the variance of the house prices compares to the variance of the amounts that the house prices differ from the values predicted by the regression line.

C. The square of the correlation is equal to the fraction of the variation in the prices that is explained by the square footage.

VI. We now turn our attention to what we do when there are more variables being used to explain a variable, in this case the selling price of the house.

A. Suppose for a collection of houses we know:

1. Age of the house in years
2. Number of bedrooms
3. Number of bathrooms
4. Distance from city center in miles
5. Number of garage parking spaces
6. Size of the lot in acres
7. Number of floors of living space in the house
8. One response variable, the selling price of the house

B. How can we deal with this more complicated situation?

VII. We do a multiple linear regression. Here's what that means.

A. *Multiple linear regression* is a technique by which we can approximate or summarize a situation where there are several explanatory variables influencing the response variable.

B. We will use the concepts that we developed for the case of paired variables, such as square footage and price, and follow the same pattern of analysis for several variables.

VIII. We know already that square footage is correlated with house price; however, if we did not already know that one or more of our variables had predictive value, we would do an analysis of variance (ANOVA), which determines such predictive value.

IX. The idea of multiple regression is that we find coefficients for each of the explanatory variables so that they combine to predict the house price.

A. The output of running a multiple regression program gives us the least squares coefficients for each of the variables.

1. Each coefficient can be interpreted as the expected amount of difference in the price of the house from increasing the explanatory variable by one.

2. For example, one more acre raises the house price, on average, by $49,200.

3. On the other hand, adding a bedroom appears to decrease the value of the house, perhaps because in given houses with the same square footage, one with fewer bedrooms has larger rooms, a feature generally associated with higher-quality houses.

B. The point is that the multiple linear regression produces a way to predict house prices if we are given values of the explanatory variables.

X. The output of a multiple regression program typically also provides additional information.

A. Among other information, the output includes an R^2 value, which tells us what fraction of the variation in the house prices is captured by this model.

B. In our case, 77.7% is captured by the model, which means that 77.7% of the variation in house prices is explained by our predictive model that used square footage, lot size, distance from center of town, and number of bedrooms.

XI. The strategy of doing a multiple regression analysis is that we find a model that predicts the response variable as a combination of the explanatory variables.

A. We measure how well the model fits by measuring how much difference there is between the predicted values and the actual data.

B. We can determine what percentage of the variation of the actual value is explained by each variable or by any set of variables.

C. Having established such a model for house prices based on all the houses that were actually sold during a year, the model might be used by the tax department to produce market valuations of all houses in the city.

Readings:

B. Bowerman, R. O'Connell, and A. Koehler, *Forecasting, Time Series, and Regression: An Applied Approach*, 4th ed., part II.

R. Dennis Cook and Sanford Weisberg, *Applied Regression Including Computing and Graphics.*

David S. Moore and George P. McCabe, *Introduction to the Practice of Statistics*, 5th ed.

Questions to Consider:

1. In our example of multiple regression, the constant term was negative. That seems to imply that if the house had 0 square feet, 0 bathrooms, etc., then it would cost some negative amount. Does that feature imply that the model is wrong? What is an interpretation of it?

2. In linear regression, the scatter plot is approximated by a line. Can we interpret the multiple regression approximation in some geometrically meaningful way?

Lecture Eighteen—Transcript
Real Estate—Accounting for Value

The previous lecture concerned war, so it seems only fitting that this lecture should concern taxes. In this lecture, we're going to confront a real-life problem and see what statistical techniques are involved in solving it. The problem that we're going to think about is producing assessments of the market values of houses in a city. For example, if you're the tax assessor, you need to give a value to every house in the city.

Some houses were sold during the preceding year, and so for those houses, we know the data of interest about those houses. We know their square footage; we know the number of bedrooms; and we also know the sales price. But for the other houses, we know the data about the houses—such as their square footage and lot size—but we do not know the sales price. So, our challenge is to give an effective prediction of the sales price of the houses that were not actually sold.

In most of our lectures so far, we've dealt with one, or sometimes two, varying quantities; but in real life, usually there are many more variables involved. If we think about what makes a house sell for a certain price, there are many features involved, including the square footage of the house, the location, the lot size, the quality, the building materials, many different features.

These real estate features are examples of variables associated with each house in the population. In this case, the population is the population of all the houses in the city or the tax district. The same basic principles that are used when dealing with one variable are used to deal with more than one variable, but everything becomes a little bit more complicated. So, that's what we're going to try to deal with today.

As usual, our goals are to organize, describe, and summarize data. In this lecture, we're going to describe methods for modeling data that have multiple variables, and we're going to accomplish this by doing a specific example.

Two basic questions of interest arise when there is more variables than one present and that is, what quantities affect or explain or are related to what other quantities? So, in other words, you know things like the square footage, the lot size, the number of bedrooms, number of bathrooms, the distance from the city center. You know that all of these things influence the market value of a house; but some of these variables might be related to each other.

For example, we would expect that the square footage of the house and the number of bedrooms or the number of bathrooms would all increase together, on average. So, the question is, "How can we summarize all these relationships?" In particular, sometimes, can we eliminate some of the variables as being not important in developing our predictive model for how much to assess a given house at? But, can we throw out some of those variables and still get a model that explains most of the price of the house, and captures most of what's going on?

Let's begin by taking our list of houses—we're imagining we have houses in the city—and we're going to take the variable that we expect to have the most influence on the value of the house. We'll start there. What variable seems likely to have the most influence? Square footage seems like a likely candidate for the most influential single variable in predicting the price of a house.

So to ground our discussion, let's look at a collection of data that consists of having the square footages, and the selling prices, and other information about a collection of 113 houses that were sold last year in our example city. Now, what we see here on this list is an excerpt from the total list of houses that were sold. You see, we're going to create a model using the houses that were sold, and then develop a model that we can then apply to the ones that were not sold. This is just an excerpt of a few of the 113 houses that were sold. You can see for each one, we have a sales price; we have the number of bedrooms; bathrooms; home square footage; lot acreage; the age of the house; and the distance from the city center. So, we have all of that information for every single house.

We decided to focus on the relationship between square footage as an *explanatory variable* to the selling price, which is then called the *response variable*. We have the explanatory variable; square footage; the response variable; and the selling price.

Naturally, we expect larger houses to sell for more—but we don't expect there to be a perfect correspondence between the selling price and the square footage. So, our statistical challenge is to describe how these two varying quantities are related and how closely they're related to one another.

Well, the first thing that we want to do is to look at these variables independently. So, specifically, let's look at a histogram of each one. This is a histogram of the square footages of our 113 houses. You can see that it has this shape where it's skewed to the right. So, this is the histogram of the square footages of the collection of houses, and this is a histogram of the house prices. We can look independently at the histogram of house prices and the histogram of the square footages, and notice that the two histograms are somewhat similar to each other. They're both skewed to the right, as expected from the idea that these two quantities—square footage and selling price—are related to each other.

We can visualize the relationship between these two variables by making a scatter plot of the two variables and, as you see on the scatter plot, for every house, we put a dot. Namely, we see what its square footage is, and then we put it at that location horizontally, and then we rise up to what its selling price was, and that's where we locate the dot for that house.

In this case, you can see that the scatter plot of these 113 points is a cloud of points that looks roughly linear and goes up to the right as expected, meaning higher square footages correspond to higher selling prices. We can approximate that scatter plot by a straight line that is the straight line that we actually met before in Lecture 7, which is called the *least squares regression line*. The least squares regression line is a summary of the data in the same sense that the mean is a summary of a collection of numbers. It's sort of a central summary around which the data are spread, either more or less. Just looking at this least squares regression line, we can notice that it fits the data fairly well.

By the way, let me remind you of what the least squares regression line means. It means that for every dot, for every house in our collection, we see what the vertical distance is between the dot and our regression line. We square that difference, and add up the squares of those differences. So, the sums of squares is a number, and we choose the straight line which minimizes that total sum of squared difference. That's the least squares regression line.

So, these two quantities—square footage and price—are related, but larger houses seem to have values that are above the line, meaning that maybe we need to think a little more carefully about those houses separately—but, nevertheless, this is still a pretty good fit.

After drawing this line, we can compute a number that captures the extent to which the data are spread out from the line. After having chosen our least squares regression line, we can actually write a formula for it. The formula for it, in this case, is exactly this: The price is equal to $161 times the square footage minus 63,600.

There are actually computer programs that will allow you to just automatically get this formula for the regression line from the data. If you look at an Excel program or any statistics program, it will just spit out this least square regression line.

In our example, then, we notice that the slope of this line is $161. In other words, the coefficient of the square footage is $161. What that means is that we expect that when we increase the square footage by 1, then this price increases by $161. So, you see that it makes sense. Roughly what this is saying is that, in this neighborhood, as a first approximation, when you add a square foot to your house, you would expect the price to increase by $161.

Now, when we look at this total picture of the houses and the scatter plot, we can think of each house in the following way. For each house, the data—the actual value of the selling price of the house—is equal to what its model tells it that it should be, plus this vertical distance, which is called the *residual*. So, the vertical distance is the amount by which the actual square footage differs from the model that we're creating.

So, let's give a particular example. If we look at this house—whose value is $649,000—if we compute its predicted value, given its square footage—which is 3,793, a big house—we can put its square footage (3,793) into our predicted formula and do the computation and get that the predicted value is $547,000.

Now, in this case, we know that the actual selling price here was $649,000. So, the residual is the difference between the actual selling price and the predicted selling price—so it's $102,000. We can see this visually on our graph—that the length of this vertical line corresponds to a difference of $102,000. Our goal, of course, is to try to get a predictive model that predicts the values as closely as possible. This one was off in this particular case by $102,000.

If we look at these histograms of the predicted house prices versus the actual house prices, we can see visually the extent to which they seem to be similar. We're trying to get a predictive model that looks exactly like the actual house prices—and you can see that it looks similar, but not quite the same.

If we graph the scatter plot of the residuals—that is to say, for each residual, how far off the model was from being correct—we get a scatter plot, and we hope that this scatter plot is a rather random-looking collection because we're hoping that there's no bias to having the residuals in one area being always higher or one area always being lower. It does look rather random. Our goal is to make the residuals as small as possible. So, our concept, again, is that Data = Model + Residuals.

When we've made a summary of data, our next step is to ask the question, "How well does our summary actually correspond to the data?" Whether it's taking just a mean or, in this case, drawing a regression line. So, what we're trying to do is to see the extent to which our model explains the actual selling price of the house.

So, here's what we're going to mean by this, very specifically. The original houses had a certain variance to them. Remember that the variance is the square of the standard deviation; the variance is just the average squared distance from the mean to each individual data point. So, the variance of the sales prices of these 13 houses is just the mean squared distance from the mean to each value of the house. So, it's a measure of how spread out the original sales prices of houses were from the mean of the sales prices of the houses.

The predicted prices gave different values for every one of the houses. You follow me? For every single house, it had a square footage, and that put the predicted value on the line, a slightly different place from its actual value. So, we could look at the variance associated with the predicted values of each of those houses.

In this case, the variance of the predicted values is this number, and the variance of the actual prices is this number. The ratio of these two—this to this—is .697. The way this is described is to say that 69.7% of the variation of the house prices is explained by the square footage. You may have heard this term, how much of some varying quantity is explained by some other quantity. How much intelligence is explained by upbringing? How much of it is explained by genetics?

The actual meaning of that term is that we take a model—in this case, a least squares regression line—and we compare the values of the model to the actual values of the selling price. We compare the variance of how spread out the original selling prices were, compared to the variance of the predicted ones, and making that comparison shows us what fraction of the model is explained.

An equivalent way to look at it is we could look at the variance of the residuals. Remember the residuals are the differences between the model predicted prices and the actual prices. We're trying to make them small. What this says is that the variance of the residuals, the leftover parts, is only 30.3% of the variance that you started with. So, this model, just using square footage, has explained 69.7% of the variation of the price.

By the way, this number—.697—is actually the square of the correlation. That was the number—the correlation, or correlation coefficient—that we discussed back in Lecture 7. Remember, it had a formula. The point is, that value of correlation, which was measuring the extent to which our scatter plot was lying close to a straight line, the square of that is actually telling us the proportion of the variation of the price that's explained by the square footage.

We now turn our attention to what we would do when we're using more variables than just that one variable in trying to explain the selling price. In other words, conceptually, for all the houses in the city, we know a lot of things about them, not just the square footage. We know the age of the house in years; we know the bedrooms; we know the number of bathrooms; we know the distance from the city center; we know the number of garage parking places; we know the size of the lot on which they sit; we know the number of floors; we know the acreage of the lot. We know all of these things about each house. So, our goal is to get the predicted value. The one response variable is the selling price of the house, and we want to use this other information to help us get a better model, a better explanation, for what the actual selling price of the house is.

So, for each house, we actually have these explanatory variables and one response variable, the selling price. How can we make use of or deal with this more complicated situation? How can we make use of this other information?

The answer is that we do something that's called a *multiple linear regression.* A *linear regression* just gave us a single straight-line summary—a predictive model—based on just one variable, like square footage. Multiple linear regression is a technique that we can use to approximate or summarize the situation where we have several explanatory variables for just one dependent variable.

The idea, though, is to use the same concepts that we've developed for the case of just a single variable—such as square footage and price—and try to follow the pattern of analysis, but apply it to this more complicated situation with many variables.

So, first of all, to begin with, we feel that the explanatory variables—like square footage, area of lot, distance to city, etc.—that these things actually do have an effect on the house price. That's the first thing. We sense that there is some significance there. But if we were starting from scratch in some other setting—I'm telling you things about the general strategy of doing multiple regression—and we didn't know whether these variables were having an effect on the actual outcome, then what we would do is something called an *analysis of variance*, or *ANOVA*. This basically tests the null hypothesis that none of the variables is a predictor of the response variable. So, that's what an ANOVA test says, that at least one of the independent variables is associated with the answer.

We would see that if we have a p value that's smaller than .05—which is typical for taking this null hypothesis threshold of significance—the result of this ANOVA test would say it would be a rare occurrence by accident to have the data we do if none of these variables were explanatory (had something to do with explaining it). Getting a small p value, by the way, wouldn't mean that the model necessarily explains a lot of the variance that we're talking about, but at least it says that it would explain some.

In our case, this step would just confirm something we already know—namely, that we have some variable that actually does make a significant contribution to predicting the house price. Because we already have seen that the square footage is correlated to the house price in a rather significant way.

The idea of multiple regression is that we're going to find coefficients for each of the dependent variables. Just like we found this coefficient of $161 for the square footage, we're going to find coefficients for each of the dependent variables so that when we combine them, they will predict the house price hopefully better than would happen with just one variable alone. That's the goal.

So, when we run this multiple regression program, it gives us least squares coefficients for the variables. So, if we take each house, we could combine the values of the explanatory variables, using the coefficients that are produced by the multiple regression, and we get a prediction of the house price. So, as before, the residual is just the difference between the actual selling price and the price that's predicted.

Let's see what actually happens if we run a multiple regression on our collection of houses. Here's what it produces. It produces an equation here. The equation is the model whose answer is the predicted price. By the way, this line and this are exactly the same thing. This is written out in words, and this is just in a more abbreviated form. So, these are exactly the same. It says the predicted price for our model is a certain coefficient times the square footage.

Notice that when we use the multiple regression model and we have more variables involved, the coefficient of the first variable, square footage, has changed. Previously, it was $161, and now it's $190 per square foot. Then, we have a coefficient—$49,200—that is the lot size, in acres. Roughly speaking, the way to think about this is that an additional acre will add $49,200 to the predicted value of the house.

This coefficient is negative, $-12{,}300 \times d$. What is d? Well, d is the distance from the city center, in miles. This makes sense because what this says is that for every mile further away from the city center you go, you subtract $12,300 from the house. You see? So, we're interpreting the coefficients of these explanatory variables in terms of a unit change in the explanatory variable has what effect on our predicted value in the model?

Here is an interesting one. In this example, –24,400 × b, which is the number of bedrooms. This one, when we see this, we have to say, "Whoa." What is going on here? Why would having an extra bedroom decrease the price of the house by $24,400 per bedroom? When we see a statistical anomaly like this—we see something that hits us and we say, "Well this doesn't seem right"—it gives us pause to try to think about it. Does it make sense, or doesn't it make sense?

Well, the answer is it does make sense. What this is saying is that if you have a house of the same square footage and you have more bedrooms, then that house is more apt to be a less luxurious house. The quality of the house is less because that means that the rooms are smaller. In the computation, it manifests itself by having a negative coefficient for the number of bedrooms. That's a rather interesting wrinkle to this thing.

Looking at the same house—you can see this is the general equation—and if we plug in the values for the exact same house that we looked at before, we see that now the predicted value is $628,000. Its actual price was $649,000—so, in this case, the residual has been reduced to $21,000. So, you see it's a better fit.

Once again, we can plot the residuals in this way to see that they're random. Of course, we're hoping that they get smaller. In this case, in fact, they do get smaller because we can measure the variance of the residuals and see how small they are. Alternatively, we can measure the variance of the predicted house prices, and compare it to the variance of the actual selling prices of the houses, in order to see what fraction of the variance in the original sales price of the house has been explained by our model. It's traditionally called R^2. It's defined to be 1 minus the variance of the residuals, divide by the variance of the actual prices. In this case, it has explained 77.7% of the variation in price. That is to say, 77.7% of the variation in price is explained by the qualities that we used in our model—square footage, lot size, distance from city center, and bedrooms—and finding the least squares multiple regression model explains 77.7% of the selling price.

This was an example to show how one can use multiple explanatory variables to try to explain one variable. Then what would happen is, once we have a good model, we would apply it to all the houses in the city to give the assessments that would be used for tax purposes. Thank you.

Lecture Nineteen
Misleading, Distorting, and Lying

Scope:

There are three kinds of lies: lies, damned lies, and statistics.

—attributed to Disraeli by Mark Twain.

In this lecture, we will learn some effective ways to lie with statistics. Lying with statistics means one of several things. We might, of course, simply present false data. But more interesting methods involve taking perfectly valid data and distorting their meaning by using misleading presentations or by drawing improper inferences. Here's one example of several we'll explore: A large college wishes to advertise that it has small classes, so it creates 99 one-student classes, then makes one class contain the remaining 901 students. Because the college has 100 classes and 1,000 students, it advertises that its average class size is 10. But 901 of the 1,000 students experience a class size of 901. The mean often does not suggest a meaningful story. By examining several misleading uses of statistics we learn to recognize the inadvertent or purposeful misuse of statistics.

Outline

I. Mark Twain attributes the following quotation to Disraeli: "There are three kinds of lies: lies, damned lies, and statistics." Techniques of analyzing and presenting statistical data can be misused, intentionally or unintentionally, to give distorted views of the world.

II. Here's a misleading statistical fact: The average American has one testicle and one ovary.

A. The statement is correct but completely misleading.

B. There is a lurking variable: sex. Bringing the lurking variable to light gives quite a different view of the data.

C. For many cases of existing data, we don't know whether or not there is a lurking variable.

III. Here's an example showing that outliers can have so large an effect on the mean that merely stating the mean gives a distorted view of the actual income distribution.

A. In a recent year, the mean increase in net worth of graduates of Lakeside High School in Seattle was more than $2,000,000.

B. But the reason is not that a lot of the graduates make millions of dollars a year.

C. The reason, instead, is that the average included Bill Gates and Paul Allen, graduates of Lakeside High School and the founders of Microsoft Corporation—extraordinarily wealthy outliers whose net worth greatly distorted the mean.

D. A much better measure of the center than the mean would be the median, which is not affected by outliers.

IV. Even something as simple as class size in a university is a bit tricky to summarize.

A. Let's do an extreme example, making the unrealistic assumption that each student at the university takes just one class.

B. Suppose there are 1,000 students, with 901 in one class and with each of the remaining 99 students individually tutored.

C. The mean class size is 10, an accurate but misleading summary.

D. The median class size is 1, which also is misleading.

E. A better summary value for expressing the average experience of class size by a student is to add the class size experienced by each student ($901 \times 901 + 99 \times 1$) and divide by 1,000. The result is 811.9, a better summary number.

F. The median "class size as experienced by the students" is 901, also a good summary.

V. Biased samples, some intentional, some not, are common.

A. Using a sample that is not representative of the whole population gives a distorted view. In an earlier lecture, we saw the example of the *Literary Digest*'s biased sample.

B. As a source of understanding our world, our friends form a biased sample.

1. The people we know, on average, tend to be like us.
2. We can easily believe that everyone thinks like us because every time we ask our friends about something, they tend to agree with us more or less.

VI. The wording of questions in a survey can influence results.

A. Surveys can either intentionally or unintentionally have questions worded in ways that affect the way people answer them.

B. Consider this question: "Would you rather have: the very risk-taking Smith, or Jones, who is likely to save us from desolation?"

VII. Virtually any news source is biased, in the sense that its contents are chosen for interest.

A. Frequently, the interest in a story comes from being rare and being bad.

B. Television news is biased toward stories with visual content.

C. In a recent year, the rate of death from terrorist attacks in Israel was .00038, which is one-third the rate of death from traffic accidents in the United States. But terrorist attacks are better news stories.

D. Any news source is biased, but we must realize that an unbiased news reporting system would be dull.

VIII. Through selective reporting, statistics can be manipulated to persuade people that a particular drug is effective, even though it is not, or to predict falsely the future of stock market prices.

IX. People can answer surveys with wrong answers.

A. Sometimes people give the answer to a question that they think the questioner would like to hear.

B. Many studies showed that child molesters were frequently molested as children.

1. Some recent studies put those data into question.
2. Child molesters may gain some advantages in the legal system by claiming to have been molested as children.

X. Graphs or phrasings can be distorted.

- **A.** To make a change that is small percentage-wise look large in a graph, omit from the graph most of the possible range of the quantity.
- **B.** Another way to have accurate graphs that are misleading is to draw them so that the height of the graphical symbol accurately represents the issue, but it is drawn to look like a three-dimensional thing; we intuitively think of the thing's volume, which of course, increases much faster than its height does.
- **C.** Consider this statement: "People who eat a particular food have a 30% higher chance of contracting a certain rare disease."
  - **1.** The rate may go up from 1 in 10,000,000 people getting the disease to 1.3 in 10,000,000.
  - **2.** Yes, this is a 30% increase, but the increase may be insignificant or it may not. The simple statement "30% increase" doesn't tell the story.
- **D.** Likewise, suppose we read that Company X has increased its profits by 50% over the last quarter. This impressive statistic is not necessarily so impressive if the profit in the preceding quarter was $1,000.

XI. Extrapolating trends mindlessly can give ridiculous conclusions.

- **A.** Some of the scare stories we hear are the result of extrapolating trends.
- **B.** Trends in economic growth or population growth are a common subject of inappropriate extrapolation.
- **C.** In sports, world-record running times cannot continue a linear trend because there are physical limits to how fast people can possibly run.

XII. Confusing correlation with causation is a big source of the misleading use of statistical information.

- **A.** People often start with a true correlation but then derive a false causal relationship from it.
- **B.** Lurking variables often underlie such misconceptions.

XIII. Statistics can be an incredibly useful tool; however, we must be cautious as consumers of statistics to avoid being taken in by the pitfalls intentionally or unintentionally included in the presentation or interpretation of statistical information.

Readings:

Darrell Huff, *How to Lie with Statistics*.

Questions to Consider:

1. Many prisoners cannot read. Is it an important insight or a logical fallacy to argue that programs that teach reading to prisoners might reduce rates of recidivism?
2. To some extent the use of statistics should come with a "buyer beware" label. Where should the ethical and legal responsibility lie with regard to the presentation of statistics in a meaningful way? For example, should misleading (but technically correct) graphs be prohibited along with literally false advertising, or not?

Lecture Nineteen—Transcript
Misleading, Distorting, and Lying

You may remember Mark Twain's quote that he attributed to Disraeli, that, "There are three kinds of lies: lies, damned lies, and statistics." In this lecture, we're going to embrace that quote, and talk about how to distort, mislead, and lie with statistics.

So, we're going to give some examples. One of the common ways to mislead with statistics is to use the mean inappropriately. One example we already talked about before was the fact that the average American has one testicle and one ovary. The mean is completely legitimate—there's no question about actually doing the computation correctly—but it's somehow misleading because it fails to mention the lurking variable of half of the population being male, and half of the population being female.

Here is another example. Suppose that you are trying to find the school that's just perfect for your child. You look around the country, and you think about various qualities, such as academic quality, but eventually you decide that it would be better simply to go for the bottom line and ask what school has graduates whose average wealth is higher than other schools. After looking around, you find a particular school where the average, the mean, of the wealth of each student is in the millions of dollars, over all graduates, over the many-decades history of the school. You say, "That's the school for my child," and you enroll your child in the Lakeside School in Seattle, Washington. Then you go to the alumni meetings, and you see people around, and you notice that, although they are probably far wealthier than average, you discover they're not all millionaires.

Then you realize the error in your ways, namely that two of the graduates of the Lakeside School were Bill Gates and Paul Allen, founders of the Microsoft Corporation. Their wealth is so vast that taking the mean over the remaining many, many hundreds of students who graduated from Lakeside School, everyone else could make zero money and, still, there would be millions of dollars for every graduate.

Here's another example that's very simple, yet can be misleading. Suppose that you're at a university and you want to advertise how much attention individual students get. You want to tell what the average class size is among all your classes. Well, what does that mean, average class size, or mean class size? One way to think about it is you simply take the number of classes and divide by the size, and then get the average.

Suppose that we have a school that has 1,000 students in it. It has 1 class with 901 students, and 99 classes that are individual tutorials. In that case, there are 100 classes, and 1 of them has 901 students, and the other 99 have 1. So, the average students per class are 10.

But, on the other hand, if you ask the students what their opinion is of the average class size, of course, 901 of them feel that they're in a school that has a class size of 901; only 99 of them feel that they're in a school that has class sizes of individual attention. So, from the point of view of the students, each of the 901 students says, "I'm in a class of 901." So, we multiply those together; add the 99 students who were in class size of 1; divide by the 1,000 students; and we see that the mean from the point of view of the students is 811.9 students per class.

In this case, the median class size, being 1, is not a very good representation either, and the median class experience of 901 may be better. Sometimes the median is a better summary than the mean. For example, if we're looking at the Lakeside School example again, the median would be a more representative number for telling what the, so to speak, medium central wealth is of the graduates of that school. Taking the mean is a good way of distorting data.

Another kind of example is bias. Bias occurs in many places in statistics. We saw in the *Literary Digest*, taking a sample of only wealthy Americans in the 1936 Presidential election, and getting a far unbalanced view of reality. But I want to talk about the kind of bias that happens to each of us every single day. That is for all of us, our existence and our impression of the world is obtained by looking around at the people who are immediately around us—namely, our friends and our families. Those are the people from whom we get an impression of the world.

Our friends and family are wonderful people, but they are not representative of the whole population. The people whom we know, in general, tend to be like us. That's why we like them. We can easily believe that everybody thinks exactly like us because every time we ask people around us, sure enough, they more or less agree with us. Our friends are wonderful people; they are delightful to have around us; but what they are not is a representative sample of reality.

Another way to get biased samples is wording of surveys that is completely biased. If somebody takes a survey and asks the following question, "Would you rather have the very risk-taking Smith as your leader, or Jones, who is likely to save us from desolation and also loves his dog?" Which one are you going to choose? Asking survey questions in ways that are either intentionally or unintentionally misleading can cause the answers to surveys to be very biased and not representative of what people really think.

One of the most serious biases of all is the bias of newspapers. I know you've all heard of the bias of newspapers. Usually when I say, "What's the bias of newspapers?" you're thinking the bias of being too liberal or being too conservative. But I'm talking about a bias that is far more fundamental than that. The real bias of newspapers is that they are interesting. Every newspaper prints stories that are interesting. That is the fundamental criterion for any story. But what does it mean to be interesting? To be interesting means that it's a rare event; that's almost the definition of interesting. It's got to be a rare event and usually, by the way, it has to be a rare event and a bad event.

If you go through the first section of a typical paper, line by line, the fact is that almost all of the articles are bad. When my daughter was 16, I was thinking it would be good to get her in the habit of reading the newspaper. So, for a couple of days, we took *The New York Times*, and I read some stories from the newspaper. After a couple of days, she said, "Daddy, I don't want to do this anymore." I said, "Why? These are interesting articles." She said, "Because everything is bad. It talks about war; it talks about criminal behavior; it talks about death." I'll actually illustrate this with numbers to illustrate to you how biased the reporting is, in that sense of choosing interesting things.

Suppose you're contemplating taking a trip to Israel. If you're thinking about going to Israel, of course, they're constantly having terrorist attacks in Israel, and people die in these attacks. You might ask yourself, "Is it safe to go to Israel?" One way to determine that is to actually look at the data.

If you look at the data, in a recent year, the data say that there were 248 people killed in this one year from terrorist attacks. The population of Israel is 6.5 million. If we compute the rate of death per person in Israel for a year, and then we do the similar computation for the rate of death from automobile accidents in the United States, we will find that the rate of death from automobile accidents in the United States is three times the rate of dying from terrorist attacks in Israel. Yet we say to ourselves, "I get in the car every day, and it doesn't bother me. That's not something I worry about." Yet going to Israel is something you would consider this. It's because every one of those 248 deaths in Israel was on the first page of the paper. This is bias.

If we had a newspaper that was not biased towards the interesting, it would be a very dull newspaper indeed. It would say, "Today, Mr. Jones did not commit murder. Today, Mr. Smith did not commit murder." It would be a very dull newspaper.

There are other issues about selective reporting, by the way, that also give misleading reports. Suppose that you have invented a miracle drug. It's a miracle drug in the sense that it really doesn't do anything. It's a complete wash; it has no actual active ingredients. But you want to sell it as a miracle drug, and you know that to be persuasive, and to sell things as a miracle drug, you need to have good statistical evidence that it's an effective drug.

Well, we know how to get such evidence. We know how to take tests because we followed this course. We know that the way to do it is to do double-blind procedure tests where you take a placebo versus the purported miracle drug; you give it to a collection of people; you don't tell them which one they're getting; you don't tell the experimenters which ones they have; we know how to design an experiment.

So, here's what you do. If you want to sell your miracle drug, all you do is realize that there's the possibility of a Type 1 error. Remember what that means. The null hypothesis, when you're testing a drug, is that the drug has no effect. A Type 1 error occurs if, in fact, it has no effect, but the evidence indicates that it does have an effect. So, if you want to sell your miracle drug, all you need is a Type 1 error. How could you get a Type 1 error? It's very simple. All you do is do several hundred experiments—perfectly designed experiments, wonderful experiments, double-blind experiments; everything is perfect.

You know that a certain percentage of the time, if your significance level is .05, as is typical—then about 1 out of 20 times, in fact, the results of the experiment will indicate that your drug is, in fact, efficacious. Actually, 2.5% because half of the time, it will be lower than the average of the others. So 2.5% of the time, the test will indicate that the drug is effective, even though it has no effect whatsoever. So this is a great way. Then all you do is you report that test. The person reads the test in the paper; they say, "This was a very well-conducted test," and they're persuaded to buy your miracle drug.

There's another example of this, and that is how to become a perfect predictor of the future of stock prices. It's very simple, actually, to be a perfect predictor of the future of stock prices. Here's what you do. You send out 1,024 predictions at the beginning of a week to 1,024 people, and you say, "I have special insight, and I will tell you whether Dell stock will go up or down next week." What you do is you send 512 people the prediction that it will go up, and you send to 512 people the prediction that it will go down. After the week is over, you see which it did.

So you've been right for 512 people—so for 512 people, your credibility is a little bit up; so to 256 of them, you send a further note saying next week, it'll go up; and to 256 of them, you send a note saying it will go down. Suppose it goes down this time. Then you send 128 up, 128 down, and so on, all the way through. Pretty soon, at the end, you have been right 10 times in a row. When you're right 10 times in a row, that's a lot of very credible evidence to the person who received those reports that you can actually predict the stock price. Then maybe you can ask them to pay you for the next week's prediction, and that person would probably pay you. That would be great.

Another way that we can be misled with selective reporting is by simple lying; people on surveys lie. For example, you have probably heard that child molesters often were molested when they were children. I'm not an expert in this—so I don't claim that this is absolutely, definitely true; but it turns out that some studies seem to indicate that, in fact, that's simply not true. It's just that child molesters realize that there's some advantage to claiming to have been molested as children in the penal justice system—they're treated more leniently in court—so they simply lie about that. Consequently, it's not 100% clear what the truth is.

Another way to lie with statistics, which is very common way of lying with statistics, is by the use of appropriately drawn graphs. Graphs are methods for displaying data, and they can be very useful, of course. But here's an example of a graph that shows an amazing trend in this particular example, where you can see that this is an example of the increase in the national debt over a month period. You can see how vast it is, you see? You can see the big numbers on the side that show how strongly it increases.

The problem is that this graph has a big distortion to it. Namely, you've chosen to make the labels on the vertical axis be spread out so much, relative to starting from zero, that it looks as though this is a huge increase—when, in fact, the reality is that the increase is only a small amount. So, by choosing the axis appropriately, you can make even a little, tiny percentage increase look vast—just by your choice of the values that you put on the y-axis.

Another wonderful example of graphical distortion comes about by drawing graphs that try to display data in one way, and actually mislead by having a three-dimensional aspect to them. Suppose that you're trying to show that Americans are becoming more overweight, and you've detected an increase. One way to display that would be to just draw a bar graph that showed a modest increase from one level to another.

But if you wanted to really drive the point home, what you could do is draw a figure—like the ones you've seen in graphs, where they try to draw them. You could draw a graph where you have a picture of a very thin person of a certain height, here, and then next to that person, you have a person whose height really does accurately indicate the data that you're trying to convey; but you draw this person as a rather large person that's sort of overweight-looking.

The effect is that visually, when we see something that has volume to it, we naturally have an intuition that it's vaster because it deals with the volume rather than just the height. So, that's an excellent way to distort data by drawing a three-dimensional histogram, when you're really just referring to the height.

Another example of ways to mislead with data is the following. Suppose that you want to scare people about the effect of a particular danger in society—whether it's a food that may be dangerous, or some other danger. One thing that you might find is that you might report a 30% increase in the risk of, say, heart attack if you eat this particular candy. It may be completely right; maybe there is a 30% increase. But what is it 30% of? Maybe your original danger from whatever it is you're talking about was 1 in 10 million. If there's a 30% increase, it increases to 1.3 in 10 million, and it may be completely insignificant. But by representing it as a 30% increase, it sounds really bad.

Or, on the good side, suppose you're in a company, a multi-national company, and your company really hasn't been doing so well. Maybe in a quarter, it just barely made a little, tiny bit of money. To exaggerate, let's say it made $1,000; this multi-million-dollar company is just barely on the positive side. Then the next quarter, it reported 50% increase in profits. So it goes up from $1,000 to $1,500. It's accurate—it's not lying; but, on the other hand, it's a misrepresentation. You get the wrong sense of the significance of that 50% increase.

A wonderful example of looking at data in different ways to get a distorted view—or not actually so much a distorted view, but just a different view, actually; I'm going to argue that these are both completely legitimate views—in this case, is the question of tax-cut savings. People often are proposing tax cuts or enacting tax cuts, and there are always two sides to the question of whether or not the tax cut is a good thing for the population in general. People draw charts to represent whether or not it's a good thing.

Here's an example of a chart that represents the effect of a particular tax cut that is proposed. The tax cut, here on this axis, it's a histogram that has ranges of incomes, and then on this axis, it has the percentage savings from the tax cut, the expected percentage savings for a person in these different ranges.

You can see that it's a rather evenhanded tax cut, that all of the people in the whole range of salaries will roughly get the same benefit from the tax cut. Even the people here who make more than $200,000, they actually get maybe a slightly smaller percentage benefit from their tax cut than some of the other people in the population.

The opposing political party will produce a different graph that really illustrates exactly the same material. Namely, its graph will say, "What is the dollar amount savings per person for the total group of people in that tax bracket?" In this way, you see a much higher graph at the end because of the fact that the total number of people who make more money—the total wealth in the higher-income brackets—is so high that the total amount of actual dollar savings would be much more for those people.

Which of these views of the tax savings is correct? The answer is they're both correct. They're both presenting actually accurate information—but one of them is trying to make a point that the wealthier people, in total, are saving more money; whereas, the other person is making the point that everybody is being treated in an equal way from the point of view of percentage savings. I am waiting for the day when a politician will show us both graphs and explain why both graphs are telling the same information, and that it's up to us to try to judge which one is the more persuasive, based on philosophical grounds.

Another category of how to distort and mislead from statistics has to do with extrapolation. Inappropriately extrapolating data is a case that can lead to some really ridiculous conclusions and can also be very scary. One example has to do with population growth trends. In the 20th century—that is, between the year 1900 and 2000—the world experienced a very fast increase in population growth. The rate of growth during that century was 1.3% per year.

One thing that we can do when we have a rate like that is to extrapolate that population growth to a much longer period of time. For example, let's just take that population growth, starting at the year 2000, and assume that we have a 1.3% growth rate—as we, in fact, experienced for the entire last century. That's 100 years; that's a lot of information that gives us a rate. Suppose that we say, "How many people will be alive in the year 3000?" Well, one of the properties of growing at a certain rate—like 1.3%—is that there is an exponential growth effect, which leads to a curve such as this one.

If we simply extrapolate to the year 3000, we will discover that this is the astronomical number of people that there would be on Earth in the year 3000. This is so astronomical—2,441,000,000,000,000 people. To show you how vast this is, if we divide that number of people by the number of square feet on Earth, we'll discover that, roughly speaking, if I did the computation correctly, basically, there will be two of us in each square foot in the year 3000. Just for your information, this will not happen.

But there are other trends that are followed. For example, here's a trend. These were data about the debt held by the public for a few years, and it actually declined in about 1998. Here's a quote from the Office of Management and Budget. It said, "In 1998, the federal budget reported its first surplus—$69 billion—since 1969. In 1999, the surplus nearly doubled to $125 billion and then again in 2000, to $236 billion. Under the President's budget proposals, $2.0 trillion in federal debt, held by the public, will be retired over the next 10 years, all of the debt that can responsibly be retired." That was written. By the way, let's see what actually happened. Ah, here are the new figures—the real figures—for the public debt. That trend simply didn't quite happen as hoped.

There are other kinds of extrapolations that we can make, like world records, for example. If we look at world records for running a mile over this last century, you can see these dots here. This is a scatter plot that shows the world record for the mile as it has occurred during this last century. You can see that it's well approximated by this straight line. You see? So, this looks like a good trend to extrapolate. Let's go ahead and extrapolate it. Here it is. We extrapolate the trend at exactly that same slope, and we discover that in the year 2600, the person will have run a mile before they even start; it will take less than zero time to run a mile. That's an example of extrapolation that may be a little bit exaggerated.

Another category of misleading with statistics comes from confusing correlation with causation. We talked a little bit before about some examples of these. One example is, suppose that you wanted to promote family values by pointing out that married people make higher salaries than unmarried people. That may be a true fact, but why? Married people may, on average, be older. In fact, it would be more persuasive to say people who have been married for 20 years make a higher salary than people who are unmarried because there's a lurking variable. There's a hidden variable that really explains the increases in salaries, and it's not really a cause-and-effect relationship.

Another example of correlation that you may detect is not actually cause and effect is IQ versus shoe size. Little babies are often cute, but their IQs and their shoe sizes don't tend to be that big. As they grow, their shoe size grows and their IQs grow—but, in fact, that's not cause and effect. It probably wouldn't be useful for my students to buy big clown shoes before the next test on statistics in order to help them out to do better on the test.

Another example of correlation versus causation is that the people who are in first class on airplanes tend to be wealthier than people in coach. So, you might think that the road to wealth is simply to buy first-class tickets whenever you fly. Of course, unfortunately, that may actually have the opposite effect.

Well, I hope you've enjoyed an excursion through how to distort, mislead, and lie with statistics.

Lecture Twenty
Social Science—Parsing Personalities

Scope:

Social policy and social sciences rely on the interpretation of statistical data. This lecture discusses two separate topics related to the application of statistics to social science. The first is a statistical technique, *factor analysis*, which can shed light on what quality several correlated, measured quantities all might be measuring. The technique seeks to identify underlying latent factors that explain correlation among a larger group of measured quantities. The other topic is possible limitations of hypothesis testing.

Outline

I. In this lecture, we will discuss two aspects of statistics that arise in the social sciences.

- **A.** The first is a technique called *factor analysis*.
- **B.** The second part of the lecture discusses an issue within the social sciences community concerning the over-reliance on hypothesis testing and its validity.

II. Factor analysis is a statistical technique that tries to find whether data comprising a number of variables can be summarized, or explained, by a smaller number of "factors."

- **A.** Charles Spearman, studying intelligence, is credited with inventing the technique of factor analysis about 100 years ago.
- **B.** He hypothesized that there is one underlying factor of general intelligence (called the *g factor*) that underlies results of various other measures of mathematical and verbal skills.

III. The assumption underlying factor analysis is that there is a small group of latent factors that accounts for the correlation among a larger group of observed variables.

- **A.** For example, a questionnaire might ask 50 questions about emotions, each with a numeric answer, such as, "How much fear are you feeling?" "How much control are you feeling?" and so on, yielding 50 observed variables.

B. The factors found in factor analysis are chosen specifically so that they have no correlation. They represent independent characteristics (of a person).

C. A successful factor analysis would yield a small number of factors that explain much of the total variation in the original data.

IV. We'll begin by looking at the famous Myers-Briggs Personality Type Indicator.

A. After answering about 100 multiple-choice questions, the Myers-Briggs test presents us with a summary of our personality or preferences using four scales: (1) extraversion/introversion, (2) sensing/intuition, (3) thinking/feeling, and (4) judging/perceiving.

B. The results show us where on each scale our answers put us.

C. A technique called factor analysis, though not used historically in the development of the Myers-Briggs indicators, describes how the 100 questions give rise to four axes.

D. The idea is that the answers to the 100 questions can be combined in specific ways to reveal a separate rating for extraversion/introversion; another combination gives the answer to thinking/feeling, and so on.

V. Finding the combinations that give uncorrelated combinations is a mathematical procedure.

A. The researcher calls each factor an evocative summary name, depending on the ingredient variables.

B. For example, in a study of jealousy, 20 measures of qualities were reduced to 3 factors, called by the researcher Reactive Jealousy, Anxious Suspicion, and Interpersonal Insecurity.

1. The correlations between each of these 3 factors and the 20 original variables served to divide the 20 into 3 groups. The variables in the first group are highly correlated to the first factor and usually have low correlation with the other two factors, and so on.

2. Part of the intent is that something has been learned about jealousy, namely, that there are 3 principal ingredients that underlie the original 20 variables.

C. Care must be taken in interpretation, but insight can be gained about the psychological or social issue being studied, as reflected in the factors that are a good mathematical model or summary of the data.

VI. We now turn to some issues concerning hypothesis testing.

A. Hypothesis testing is an example of a strategy for testing features of our world, but some social scientists and many others feel that there has traditionally been too heavy a reliance on hypothesis testing as a way of adducing and interpreting evidence.

B. Another model of progressing toward a clearer understanding of our world may be closer to the way we often proceed in real life.

1. How do we come to evaluate people in our world? We have a sense of a range of what they might be like, but we don't know.
2. Then, we update our view. This strategy underlies a statistical point of view known as *Bayesian statistics*.

C. Perhaps the best way to understand the distinction between this updating model and standard hypothesis testing is to examine the following three experiments—all hypothesis-testing experiments with statistically identical results.

1. In the first experiment, a musicologist is presented with 10 pairs of sheet music, and he must determine if the composer is Mozart or Haydn. He is correct 10 times out of 10.
2. We are already familiar with the second experiment of the lady tasting tea, in which a lady was able to taste 10 pairs of cups of tea with milk and tell correctly each time whether the milk or the tea had been poured in first.
3. In the third experiment, a drunk claims he can tell whether a coin will land heads or tails every time it is flipped. And, indeed, he does guess correctly 10 times out of 10.

D. Most people would say that the Mozart/Haydn experiment was very persuasive and the lady tasting tea less so. With the drunk person, we retain a good deal of skepticism. We are updating the prior assessment of our view of reality.

E. In the Bayesian point of view, we view our assessment of the world as a probability graph rather than a fixed number. It is an interesting philosophical perspective that takes our everyday experience and captures it on a mathematically sound footing.

Readings:

Vic Barnett, *Comparative Statistical Inference*.

Donald A. Berry, *Statistics: A Bayesian Perspective*.

Donald A. Berry and Bernard W. Lindgren, *Statistics: Theory and Methods*, 2nd ed.

David S. Moore and George P. McCabe, *Introduction to the Practice of Statistics*, 5th ed.

B. K. Gehl and D. Watson, *Defining the Structure of Jealousy through Factor Analysis* (available at http://www.psychology.uiowa.edu/students/gehl/definingjealousy.doc).

Questions to Consider:

1. What combination of environmental and genetic factors most influences a person's desire to learn? What data would you seek to specify such combinations?
2. Suppose you flip a coin and cover it up before seeing how it landed. Do you believe it is meaningful to say, "There is a 50% chance it is heads?" or do you feel that because it is either heads or it's not, an assertion about its "headedness" is not susceptible to a probabilistic analysis?

Lecture Twenty—Transcript
Social Science—Parsing Personalities

Welcome back to *Meaning from Data: Statistics Made Clear*. In this lecture, we'll discuss two statistical issues that arise from the social sciences. One of them is a statistical technique that's called *factor analysis*. It's a technique by which a statistical model of a complex situation leads to underlying conceptual organization of the phenomenon that we're discussing.

The second item is that some social scientists feel that there's an over-reliance on a rather formulaic application of hypothesis testing to their field. So, in the second part of the lecture, we'll discuss the *Bayesian* view of statistical reasoning, which in some sense, presents an alternative take on hypothesis testing and a different view of how we can look at the world and how statistical evidence changes our mind about what the world seems like.

We'll begin with factor analysis. Factor analysis is a statistical technique that tries to find whether data comprising a number of variables can be summarized or explained by a small number of factors. Additionally, we can get insight about the field by interpreting these factors, which come about in a mathematical way that we can interpret them as explanatory concepts. We'll see examples of that to show what we mean.

So, factor analysis is a way of creating concepts, and it creates concepts that sometimes can be helpful in the description or analysis of phenomena in the social sciences. It was Charles Spearman who's credited with inventing the technique of factor analysis about 100 years ago. He was studying intelligence, and he hypothesized that there was one underlying factor of general intelligence—he called it the "g" factor—that would underlie the results of all sorts of different measures of mathematical and verbal skills. The term "factor analysis" was actually introduced by Thurstone in 1931.

The assumption that underlies factor analysis is that there's a small group of latent factors that account for the correlation among a much larger group of observed variables. I know this is very abstract now, and we'll bring it down to some examples in a minute.

Let's think of an example. Suppose you have a questionnaire that asks 50 questions about, say, emotion. Each answer has some numerical value. One question might be, "How much fear are you feeling?" or "How much control are you feeling?" Well, those are the kinds of questions. After the results are gotten, each person has 50 numbers that represent the results of that individual. So, there are 50 observed variables, and if you give this instrument to 1,000 or more people then you have these 50 variables from each person.

So, the results for any two of the questions may be correlated to each other in the population. You may find that most of the people who answer Question 3 with a strong, positive answer will also answer Question 17 with a strong, positive answer. For example, it may be that having more fear may somewhat correlate with having less control; they're correlated to each other.

Factor analysis is a way of looking for these underlying factors and measuring how much of the variation of the observations—that is, the survey results from all the people in the population—is explained by focusing on just these factors.

A factor is not just, say, one question or another question, although it could be that. A factor generally is a linear combination of the answers to a collection of the observed variables. What I mean by that is that you might have, for example, a factor be the answer to the fear question minus 2 times the answer to the control question, and that combination would be a factor. Generally speaking, if we're talking about 50 things, a factor would include many, many combinations involving many of the answers to many of the questions, maybe 20 or 30 of the questions, or maybe all 50 of the questions. So, a person's value for each factor is computable from the values that that person answered from the original variables.

One of the features that you want from a factor analysis is that the different factors should be uncorrelated with each other. In other words, if you isolate a factor that is capturing some aspect of that personality, you want the other factors to vary independently over the population.

So, in particular, we don't want any correlation between the answers to the computation of one factor and the computation of another factor. They're like the x-axis and y-axis in a graph. If you change the x value, you don't necessarily change the y value. The goal is to find independent characteristics of a person that are being used to describe the phenomenon.

So, a successful factor analysis would yield a small number of factors that explain a lot of the variation we see in the population about whatever it is that we're discussing—and yet, there are just a few factors that explain all of it. So, if you know the values for the factors, you can get a good estimate of the values of the original variables. So, that's another characteristic you want. The goal of factor analysis is to summarize complex data—sort of analogous to how linear regression was a way of summarizing data.

Let's ground our discussion with an example that perhaps will make it more reasonable. We'll begin by looking at the famous Myers-Briggs Personality Type Indicator. Most of you have probably heard of the Myers-Briggs test. Over 2 million people a year take this test. It was created originally to capture Carl Jung's personality type concept. He had a concept of personality types, and the designers of the Myers-Briggs tests were trying to capture Jung's concept into this practical form.

It's a little bit ironic, by the way, to use something that came from Jung in this lecture on statistics. He has a quote about statistics where he said, "You can prove anything with statistics." He was not necessarily a great fan of statistics as being the right way to come to insight into human behavior.

Nevertheless, the Myers-Briggs test is an instrument that tries to categorize personalities. Here's the way the test works. After you've answered about 100 multiple choice questions, the Myers-Briggs test takes those answers and then presents you with a summary of your personality or preferences along four scales, and these scales are as follows: the scale of extraversion/introversion; the scale of sensing/intuition; a scale called thinking/feeling; and a scale called judging/perceiving.

The output of taking the test is that you get a score for each of these scales—how extroverted or introverted you are, where you are on this line; thinking versus feeling, whether you tend to think carefully about things or rely on your feeling about them; and so on. These are four independent axes.

The concept of this instrument is that it summarizes a very complex thing. The ultimate goal is to summarize your personality type by using just these four scales. But, in particular, the actual information that it's taking are the answers to these 100-plus questions, which are then used to compute your answers on each of these scales. So, the point is that these are factors that summarize the much more complicated situation of the answers to all these questions.

So, for example, we have 100 questions, and maybe the questions that are associated with the extraversion/introversion factor would be a combination of the answer to Question 1, plus 3 times the answer to Question 2, minus 4 times Question 5, and so on. A specific combination of those questions (of the whole 100 questions that you asked) is combined to put you on that one scale.

A good factor analysis has the property that the factors are independent of each other, meaning that if we look at all of the people who take this test—and as I say, 2 million people a year take this test—we have a lot of data about people who take this test. In order to have a good factor analysis result, what you would want to have is independence about the extraversion/introversion result compared to the sensing/intuition result.

In other words, if we plotted a scatter plot of people in the population, and for each one, we put down their score on the introversion/extraversion scale, and then we put down their score on the thinking/feeling scale, you see every person would have a dot on this scatter plot—of these, I've just picked two of the four factors—and we would find that these results are uncorrelated. That's one of the features of factor analysis, that you would want the results to be uncorrelated, because you want each of the factors to be giving you an independent concept about what it is that you're measuring.

The Myers-Briggs test was created, as I said, to capture Carl Jung's concept of personality types, so the factors were not a result of a mathematical analysis. However, they have been confirmed by a mathematical analysis. That is to say they are basically independent, and there isn't correlation among the factors—so they do capture what it is they're trying to capture. But the factor analysis in the case of the Myers-Briggs scale was an after-the-fact concept.

On the other hand, sometimes you can use factor analysis as the way to gain insight about the field under study by first looking at the original variables and combining them into factors that are obtained from a mathematical process, and seeing whether the suite of questions seems to suggest that that quality that came from this mathematical analysis really is something of significance in the world.

To give you a little more insight about what the mathematics is involved, let's do an example. Let's say you take all of the factors, all of the variables that you originally measure—in other words, you're given these 100 questions in the Myers-Briggs test, so you actually have 100 different collections of values—and you have a whole population of people who take this test. For every two questions in the Myers-Briggs, you could ask, "To what extent are the answers correlated among the population or not correlated?" You see?

So, you can make a correlation matrix where you have every variable on one side, and every variable on the other side, and for every point in that rectangular array, you have a correlation measured as we've seen before—using the correlation coefficient—and you get a number. That is a matrix that operates on a many-dimensional space. The mathematical side of doing this factor analysis is a question of doing some linear algebra that's associated with trying to diagonalize a matrix, and doing some mathematical themes whose effect is to produce this linear combination of the various collections of variables so that you have one collection that is not correlated to the collection that's representing the next scale that is created by the factor analysis.

Let's give an example where the factor analysis itself created an idea. Finding the combinations that give uncorrelated combinations is a mathematical procedure. The mathematical result of factor analysis, then, sometimes gives insight about the field under study by looking at the original variables that are being combined into the factors and seeing whether that suite of questions that came out from the math seems to suggest a quality that should be identified and explored.

A researcher, then, would call each factor by some evocative summary name, depending on the ingredient variables—trying to say what that combination is capturing. For example, in a study on jealousy, many measures of qualities about jealousy were measured, and people were given survey questions from several surveys, such as an interpersonal jealousy scale and a romantic relationship scale. From all of these surveys, 20 variables were extracted. Factor analysis was done on the result, and three factors were identified.

These factors, then, are combinations of all of those things that were measured, but the researcher looks at one of those combinations and says, "That collection of qualities that are measured by these different questions seems to be capturing something. I'll give it a name." So, in this case, the researcher called one of the factors "Reactive Jealousy"; another one, "Anxious Suspicion"; another one, "Interpersonal Insecurity."

The correlations between each of the three factors and the 20 original variables divided the 20 into these three groups. That was the point of the factor analysis. But the variables in the first group are highly correlated to the first factor, but have low correlation with the other factors. In other words, you could have a person who was high on one of these three scales and low on the other ones, or high on both; they move independently from each other. So, the names of the three factors then reflect the researcher's interpretation of the meaning of the underlying factors.

So, part of the intent that something has been learned about jealousy is that the three principle ingredients that underlie the 20 variables somehow explain the concept of jealousy. It has identified a psychological feature of people. Of course, care has to be taken in this interpretation, but there is insight that is to be gained about the psychological or social issue that's being studied by seeing what's reflected by the factors that came up from this mathematical analysis.

So, one of the interesting features of factor analysis is that the statistical and mathematical technique is actually valuable in identifying and isolating new ideas. So, in a sense, the mathematics points the researcher in a certain direction, and then the researcher is challenged to see whether that combination of qualities somehow is a core idea in the subject.

Let's move on to the second topic for this lecture. Einstein once said, "Mathematics is merely a refinement of common sense." In fact, that's right. The strategies of mathematics and statistics ultimately arise from just looking at the world, and taking our common way of interpreting and organizing it, and then refining those methods into some sort of a system. Hypothesis testing, which we've talked about, is an example of a strategy for testing features of our world.

But actually, in our everyday experience, it suggests that hypothesis testing may not capture all of the common methods by which we evaluate and judge our world. Some social scientists—and many others also—feel that there has traditionally been too heavy a reliance on hypothesis testing as the way of interpreting evidence. So, there's another model that is progressing toward a clearer understanding of the world that may actually be closer to the way that we actually live our lives.

Let me give an example. Suppose we meet somebody, and we don't know anything about them at all. Then, we're open-minded about what it is that they're like. We don't know where on the scale of honesty they are, or how kind they are. We have a sense that there's a range of what they may be like, but we don't know where they fit on that range. What happens? As we get to know them, we see them in action; we see them do something; we talk to them; we hear them. Then what happens? We update our view of what they're probably like.

So, this idea is that we start off with some opinion—maybe a pretty bland opinion—but then we get some data—for example, we see this person being particularly kind—and so we update our probability, and we say that person probably is kind. We see the person doing something else kind, then we say it's more likely that person's kind. Then we see the person do something awful, and then we change our minds again. It's a continuous process of updating our view of what might be.

Well, perhaps the best way to understand the distinction between this sort of updating model and the standard hypothesis testing is to look at the following example. I'm going to give you an example that involves three instances in which we do hypothesis testing. We're going to do three experiments—talk about three experiments—that, statistically speaking, are identical to each other. In other words, from the point of view of a hypothesis test, they're identical, but my question to you will be, what is our reaction to these three experiments?

Here's experiment Number One. In this experiment, there's a person who you are told is an expert musicologist. And you do the following experiment. You say this person's an expert musicologist, so you take two pieces of music—one written by Haydn and one written by Mozart—and you hand these pieces of music to the supposed expert, without saying who composed which one, and then the person says which one was composed by Mozart. Then this process is done again. You take another pair of pages of music—one composed by Mozart, one composed by Haydn—and you switch them around and give them to the musicologist and, once again, that person makes the deduction. You do this 10 times, and all 10 times, the person gets it right. Okay? Now, you feel that your sense about that is, first of all, you're not surprised, and you're pretty confident that this person is, in fact, able to distinguish Mozart from Haydn. Okay? That's a pretty persuasive experiment.

Let's look at another experiment that we actually talked about before, which was the lady tasting tea experiment. In the lady tasting tea experiment, the lady was given 10 pairs of cups of tea—one where the milk had been poured into the tea, and one where the tea had been poured into the milk. Each time she was given a pair of these cups, and she was asked to tell which one was which, and she, in fact, did it successfully all 10 times. How are we going to react to this experiment? I'll give you one more, and then we'll talk about all three.

The third example has exactly the same structure to it. Suppose that a drunk person comes into a bar and says, "I can predict the way a coin is going to land." You say, "Okay. We'll test it out. I'm going to flip the coin in the air, and you tell me how it lands." So the drunk person says, "Okay." You flip the coin and before it lands, the drunk person says, "heads," and it was heads. You do this 10 times, and he gets it right 10 times in a row.

Well, how do we respond to these three scenarios? In the case of Mozart/Haydn, we're very strongly convinced that that person is indeed able to distinguish Haydn from Mozart. In the case of the lady tasting tea, we may retain a little skepticism, but we're pretty persuaded. Still, we may have some doubts because we can't really think of what she could possibly be detecting.

But, in the case of the drunk, it's pretty clear that we still have a lot of skepticism. We know it's going to happen 1 in 1,000 times, roughly, that a person will get it right just by random luck, but we're not really convinced by this experiment that, in fact, this person can predict the future; we're still skeptical.

The point of these experiments is that all three of them had the same hypothesis-testing structure. The null hypothesis in the first case was that the person could not distinguish Mozart from Haydn. Then the result was significant; it had a low p value; it would only happen 1 out of 1,000 times by luck alone that that person would be able to detect a difference 10 times in a row.

Likewise, with the lady tasting tea, the null hypothesis was that she could not tell the difference—and yet, she could and, therefore, we were persuaded. The null hypothesis for the drunk guy was the same thing. He couldn't tell the difference, but he got 10 of them right. Yet our response to these is different and legitimately so. It seems different.

The Bayesian point of view is to acknowledge this reality, that this kind of evidence really updates a prior assessment of what we view the world to be. So, let me give you an example to illustrate the concept of the Bayesian point of view of the world.

Suppose you go into a magic shop, and from the magic shop you select a coin. You know it's a magic shop, and you know that the magic shop sells coins that have every different probability of landing heads up. In other words, some of the coins always land heads up; some of the coins never land heads up, always tails up; and for every place in-between. It's a magic shop.

Because we know where this coin came from, we're absolutely agnostic about our predisposition of thinking what the probability this coin actually has of landing heads up. We're assuming that the coin does have some probability associated with it. If you flipped it thousands and thousands of times, some fraction of the time, it would land heads up. That's what we call the probability associated with this particular coin, but when we take this coin out of the magic shop, we don't know.

So, the picture that we have, then, is that our *a priori* concept of the world is that we don't know what the probability of its landing heads will be. It could be anything from 0 to 1. How are we going to change our view of what the probability is? Well, we do some experiments. We simply flip the coin some times and take data.

Suppose that we flip the coin 4 times, and 3 times the coin lands heads and 1 time it lands tails. Then, from the Bayesian point of view, we get a new graph about what the probability is for this coin landing heads. One thing for sure is, since we got 3 heads and 1 tails, we know that it is not a 0 probability coin of getting heads because we actually got some heads.

But, let me show you how to read this graph. What this says is that there is some chance that the coin only has a .4 chance of landing heads, and yet we got this data. But there's not too much of a chance. On the other hand, there's more of a chance that the probability is higher. The way to assess this is the following. You say for every possible probability of the coin—for example, it might be a fair coin with a 50% chance of landing heads—under that assumption, we could compute what the probability is of getting 3 heads and 1 tails in 4 flips. We can mark that probability here. But notice that if we have a coin whose probability of landing heads is 70%, then it's more apt for us to get 3 heads and 1 tails, and so we compute what that probability is, and we mark it here. So, for every possible probability between 0 and 1, we compute what the likelihood would be of the outcome of the experiment if that were the probability.

So, our view of the world is that this entire distribution of possible world states is our actual belief system associated with this coin. When we get more data then, what we do is we update our view of the way the world is. When we got an extra tails, it shifted that way.

The Bayesian concept of the world is that instead of saying the world is exactly one way or it isn't, instead, we say the world has all ranges of possibilities—like the coin has a range of possible ways that it could land—and what we're updating is our sense of the probability that the world is one way or another, and that experiments then help us to update our probability. This is a little different interpretation, but we'll see an example in a later lecture where this comes up.

I look forward to seeing you next time.

Lecture Twenty-One
Quack Medicine, Good Hospitals, and Dieting

Scope:

We make decisions every day about our personal health. If our cholesterol is above some number, our doctor is likely to suggest a medication to supplement healthful eating and exercise. Such a recommendation is commonly based on statistical results of studies that show a correlation between high cholesterol and increased risk of heart attack. Aspects to consider in applying results of such a study to ourselves include how like us the people in the study are and the difference between correlation and causation. Another statistical concept, *regression to the mean*, explains why quack medicine may appear to work often.

Note: References to specific diseases and treatments and all data and interpretations are used solely for the purpose of illustrating ideas of statistics and are in no way designed to be used as medical references for the diagnosis or treatment of medical illnesses or trauma.

Outline

I. Every day, we make decisions about our personal health.

  A. Our state of health is measured by data about us (our weight, our cholesterol, levels of various chemicals in our blood).

  B. Part of our decision-making about such questions as whether to take cholesterol-lowering medication is done by comparing our numbers to those reported in studies that have been conducted with a large number of people.

    1. We may need to evaluate studies to see if they reveal significant and important information or if they yield only statistically significant information, meaning a difference is detectable, but not important.

    2. As well, the studies frequently involve many people who in many ways, such as age, weight, inheritance, or gender, are not like us.

II. What we would really like is a study involving people who are as much like us as possible, because their experience with, say, a medication is more apt to be similar to ours.

A. We'd like to *condition* the overall study data on several variables, looking at the subset of the study data that matches us with respect to those variables.

   1. In general, the concept of *conditioning* the data on some criteria means that we look at only the data with a certain feature.

   2. We prefer to respond to the statistical results conducted on people most similar to us so that we would have a sense that the studies pertained most specifically to us.

B. An intriguing possibility is to use the computational capabilities of modern technology to find people who were in the scientific studies and whose characteristics are similar to our own.

   1. We'd then perform statistical analysis on that subpopulation for, say, the effect of a cholesterol-lowering drug.

   2. With that approach, we might get rather different statistical results than results based on the larger population.

C. Tailoring the statistics to our individual personal health situation could conceivably provide a greater increase in personal health than would improvements in treatments themselves.

   1. Of course, doctors are aware that different patients react differently, and to some extent, they try to tailor their treatments to the individual.

   2. Using their experience to modify the results of the studies can be good, but it can also be problematic, because an individual doctor sees many fewer patients than there are in some studies.

III. Another statistical issue arises when we need to decide which of two hospitals to go to for heart surgery.

A. Suppose we have the following chart summarizing how successful each hospital is with each of three subcategories of patients: those entering in fair condition, in serious condition, and in critical condition.

Patient Condition	Hospital A	Hospital B	Survivors from A/%	Survivors from B/%
Fair	700	100	600/86%	90/90%
Serious	200	200	100/50%	150/75%
Critical	100	700	10/10%	300/43%
Total	1,000	1,000	710/71%	540/54%

B. Looking at the data broken down in this way, we see that B has a higher success rate in all three categories of difficulties.

C. When averaged all together, however, the impression is that hospital A is superior, with a 71% survival rate. But hospital B is superior in each of the three categories.

D. Notice that mathematically, this is the same kind of example as one we saw in an earlier lecture on gender discrimination. This is Simpson's Paradox.

IV. Another statistical question arises when trying to distinguish between a quack medicine and a real medicine.

A. The reason that quack medicines sometimes seem to work is that usually people recover from a minor disease even without medicine.

B. Quack medicines appear to work because of the phenomenon called *regression to the mean*: An ill person usually expects to return to his or her mean health situation.

V. A similar example arises in childrearing.

A. Suppose (somewhat tongue-in-cheek perhaps) that what you say to your child has no effect on the child's behavior.

B. Under that supposition, after praise or punishment, the child will usually return to his or her average behavior (this follows from the definition of *average*).

C. It will appear that punishment works (because the child who was misbehaving will usually improve), but praise has the opposite of its intended effect (because the child who just did something extra good and got praise will usually return to his or her average behavior).

D. This is similar to quack medicine; both are examples where something appears to have an effect, but the explanation is really regression to the mean.

E. Other examples of regression to the mean are athletes doing poorly after appearing on the cover of *Sports Illustrated*, tall people having children somewhat shorter than themselves, and short people having children somewhat taller than themselves.

VI. A common issue in the realm of personal health management concerns dieting and weight.

A. Many quantities, including weight, have an innate variability.

1. If you weigh yourself daily and chart the results, you'll notice that the readings can differ by a pound or two, even if you didn't "gain" or "lose" weight.

2. If you weigh yourself to the nearest half a pound, then make a histogram over the readings, you get a somewhat normal-shaped picture distributed around a central value.

B. A weight chart is a *time series*, a value for different times.

C. In the case of a person losing weight, the weight chart shows a downward trend, but it does not always go down uniformly each day.

D. We can summarize the chart by drawing a straight line, the least squares regression line.

E. The weight loss data illustrate how we use a mathematical model (the straight line) that captures the spirit and look of the data to summarize the data.

VII. Several statistical issues are related to everyday health issues.

A. We discussed the notion of conditioning on subpopulations that match a person of interest, which can give different results than analysis over the whole population.

B. We illustrated Simpson's Paradox by the example of choosing a hospital.

C. We illustrated how regression to the mean can be the explanation for quack medicine seeming to work and for punishment to work but praise not to work.

D. We saw how to summarize a time series by a straight line.

Readings:

David S. Moore and George P. McCabe, *Introduction to the Practice of Statistics*, 5th ed.

Ann E. Watkins, Richard L. Scheaffer, and George W. Cobb, *Statistics in Action: Understanding a World of Data.*

Questions to Consider:

1. Suppose you have not gained or lost weight in many months. Suppose you notice that for a period of a week, your daily weight is always less than your traditional average. Can you conclude that you have lost weight? Suppose a month passes in which you are always below your traditional average. Where is the cut-off line at which you will assert that you actually weigh less?

2. What data about a medicine would you most like to know before taking the medicine? Your answer might include rates of spontaneous recovery, seriousness of the disease, measurement variation to determine that you have the disease, characteristics of the treatment studies, or other values.

Lecture Twenty-One—Transcript
Quack Medicine, Good Hospitals, and Dieting

Welcome back to *Meaning from Data: Statistics Made Clear*. Every day, we make decisions about our personal health; it's one of the most important things that we think about. Our state of health is often measured by data. The kind of data that we are interested in are things like our weight, our blood pressure, the levels of our cholesterol, and other various chemicals in our blood stream. These data present to us a picture of our well-being.

So, the kinds of questions that we're constantly asking ourselves about how to treat our own physical condition are questions such as, should we take vitamins? We don't want to get rickets. Scurvy is bad; we should probably take a lot of Vitamin C. Should we take cholesterol-lowering medication? Should we take aspirin to reduce the chances of heart attack? Or should we take Dr. Quack's Super Snake Oil when we're just not feeling quite up to par? These are the kinds of things that we need to make decisions about.

Of course, a lot of the ways we make decisions in reality are that we use anecdotal information. That's one way. But, sometimes our decisions are based on studies that have been conducted, and the way that we use the results of studies are that typically we read about a study and we compare the numbers that we have—our measurements of ourselves—to the numbers that were obtained in some study that involved a large number of people, and then that's what we use to make a decision.

For example, suppose that a study showed that on average, people who have a certain cholesterol level that's higher than a certain number suffer more heart attacks than people who have a lower such number. If our cholesterol number is higher than that bad number, then our doctor may very well suggest to us that we take some cholesterol-lowering medication. Of course, the doctor will also tell us that we should eat more healthfully and exercise more. The doctor will always say that.

What I would like to do is to look at a chart involving this issue of interpreting data about cholesterol in order to make a decision and discuss how it is that such decisions might well be made. Suppose that the data from this study about cholesterol indicates a chart such as this. Of course, this is just a summary; this is not a complete picture of the kinds of data one would get about cholesterol or other studies. This is just for the purposes of understanding, of illustrating a statistical principle, not for getting medical advice.

Here is the kind of a chart, though, that might come about through a study about the levels of cholesterol. The study might involve 1,000 people—the sum of these numbers is 1,000—of which perhaps 500 of them have a low cholesterol reading overall and another 500 of them have a high cholesterol reading overall.

It may be that there are other more elaborate tests that allow a doctor to evaluate a risk factor for heart disease. For example, a sonogram or injecting a radioactive dye into the blood system in order to detect whether or not plaque is building up on the inner parts of the arteries. So, using other kinds of tests, it may be that the doctor can evaluate the risk level for the different people in the study. The kind of result would be indicated in this sort of a chart.

Namely, for the people who are in the low cholesterol group—below some sort of a threshold—it might be that 400 of the 500 are at low risk for heart disease, and 100 are at high risk for heart disease, even though they do have low cholesterol. On the other hand, for the high cholesterol group, perhaps it's the opposite, that 100 are in the low-risk group for coronary disease, and 400 are in the higher-risk group. This is the kind of chart from which a person needs to make a medical decision.

But actually, in evaluating these studies, we actually have to ask several additional questions. For example, how much higher is the rate of heart problems in the group that is indicated as having a higher risk? In other words, is it significant in the sense of being important, or is it just significant in the sense of being statistically significant—meaning that there's a detectable difference in the rate of heart disease, but it may not actually be an important difference? It may be too small to be important, even though it is detectable and real. That's one question we need to confront.

Another question that we need to confront is the question, "How are the various effects distributed over the spectrum of people with different cholesterol levels?" Meaning, are people who are just over the threshold line of high cholesterol, are they are much greater risk than those with lower cholesterol, or are they really about the same risk as people with lower cholesterol, and it's only people with very high levels of cholesterol that have significantly higher risks? Or, alternatively, maybe the risk is fairly evenly distributed for people with the higher levels of cholesterol, and there's sort of a threshold issue, and having more doesn't make it much worse. So, those are the kinds of questions that we would want to know in interpreting the data of the studies.

But, there are other features that we'd like to know about the studies. Specifically, frequently, the studies involve a lot of people who are, in many ways, not like us. They may have, say, different weight relative to their optimal weight than we do; or, they may be of a different gender or a different race; it may be that their inheritance is difference—that is, their parents may have had different levels of cholesterols and different profiles than our parents did.

What we would really like is to have a study involving people who are as much like we are as possible. The reason for that is the people would be more apt to react to the interventions that are suggested or have the distribution of calamities that are portrayed. Those responses are more apt to be similar to how we respond, if the people involved in the studies are more like we are.

So, let's do a specific little study associated with our cholesterol one. Let's look at a specific similarity that might be important in evaluating these cholesterol levels.

One thing that we might like to know is how the study evaluates people who have the same HDL versus LDL profile as ours. Remember, HDL stands for high-density lipoprotein, and LDL stands for low-density lipoprotein. LDL is the bad cholesterol, and HDL is the good cholesterol.

So, what we would like to do is to *condition* the overall study data on several variables, looking at the subset of the study data that match us with respect to those variables. "Conditioning" the data means just looking at those people who have the same condition as we do with respect to one or more of the different variables in the population.

Let's see what we might see if we look at the overall cholesterol risk analysis, but we condition it on the levels of HDL to LDL ratio. In other words, we look just at those people in the population of the study who had, for example, a high HDL to LDL ratio.

First of all, for purposes of illustration, I just assumed that half of the people in the study had this higher HDL to LDL ratio. Notice what happens if we compare the data from this study, conditioned on the high HDL to LDL ratio—in other words, having a lot of the good cholesterol compared to the bad.

Notice what happened. For the low cholesterol—this is the low overall cholesterol group—the ratio of low-risk to high-risk people, which previously was 400:100, remained the same. The ratio was the same, 200:50. Whereas, in the high-cholesterol group, notice that the risk factor altered in proportion. Namely, notice that originally, there was a 4:1 ratio of high risk to low risk—whereas, in the case of just looking at the people in the population who have a lot of the better cholesterol, we find that the better cholesterol has somewhat mitigated the risk and left all of these 100 people who were in the low-risk area still having low risk. A lot of people who previously were at high risk—having an overall high cholesterol level—no longer are viewed as having a high risk for having arterial disease.

The point is that if we personally had knowledge of our cholesterol and happened to have this high level of the HDL cholesterol, then this is the kind of a study that we would prefer to look at in order to make a medical decision about whether to intervene.

So, the point is that we'd have a sense that the studies pertained specifically to us because they involved people who were like us in the important ways, in ways that we think are important to making this decision.

So, an intriguing possibility is to use the computational capabilities of modern computers to find people in scientific studies on which we could determine the efficacy of medicines, where we just choose people in the studies whose characteristics are similar to our own. In other words, we would have the raw data of the statistical study and then perform a personal statistical analysis on that subpopulation.

For example, on the effect of cholesterol-lowering drugs, we would tailor the study to our individual needs. As we saw in the case of the HDL result, we might get a rather different statistical impression of the results if we based it on that subpopulation, rather than on the whole population.

First of all, I want to say a couple things about this. In the case of HDL, it's standard practice now to use the HDL ratios as part of the discussion. But, on the other hand, suppose that you're a child of parents who lived long and healthy lives, but had high cholesterol. Maybe an analysis restricting to just those people would show a different concept about whether it's prudent to take cholesterol-lowering drugs. We don't know.

So, the point of this discussion is that conditioning data on a particular subpopulation is an interesting possibility. Let me repeat the caveat, that all references to specific diseases and treatments and all data and interpretations are not to be viewed as medical advice. They are not correct. These were made-up numbers for the purpose of illustrating a point, not for giving advice about cholesterol or anything else. The goal here is just to think about statistical ideas.

It seems to me that the possibility of tailoring statistics to individual personal health situations could conceivably provide a greater increase in personal health benefits in our society than would improvements in the treatments themselves. We know for sure that some ways that we deal with medications are simply wrong. For example, if you take a bottle of aspirin, and you look at the dosage recommendation on the bottle—it says two aspirin or something—it completely ignores whether the person is a 300-pound football player who's in his 20s, or an 80-pound 100-year-old in a nursing home. Surely, we can't really believe that the dosage for those two people should be the same; but, of course people have to make these sort of generalizations.

I admit, in hospitals, sometimes the dosages are varied, depending on the weight or, in fact, I think more the surface area of the patient. Of course, people respond to medicines in different ways. Some people respond very dramatically to little doses, and some people don't, and each of us lies somewhere on a continuum of responsiveness to medications and determining the right amount might be quite a different enterprise than the way that dosages are currently determined.

So, tailored statistics might lead to an incredible increase in personal health improvements. It brings up the prospect that maybe each of us will have to hire a private statistician to take the raw data from scientific studies and analyze that data to make the best inferences that apply to us. Maybe it'll be part of the doctors' procedures to look at such data. You can easily imagine that. Of course, doctors already are aware of how individual patients react. That's one thing, when you get to know a doctor, to some extent, they do try to tailor their treatments to the individual, and they use their own experience and the details of an individual's case to modify the treatment. Of course, that's good (using the doctor's experience), but it also is somewhat problematic because an individual doctor sees many fewer patients than there are in good statistical studies, and so it may be biased by small sample issues that we've seen before.

Let's look at another example associated with health decisions, which is the following. Suppose we need to make a decision about which hospital to use, and we have some data about a hospital, and we're about to enter the hospital for some serious operation. So, we decide to be prudent about this and look at the facts, and so we go to some database, and we see that there are two hospitals in our city that we could consider for attending to this serious problem. We notice that these are the results.

In Hospital A, of the last 1,000 patients who entered the hospital for this disease, 71% survived, and Hospital B, only 54% survive. We might say it certainly seems clear, from these data, that Hospital A is the preferred hospital. Let's go to Hospital A.

But, wait a minute. Let's suppose that we look at the data broken down a little more specifically into subgroups. Here we go. Here are three subgroups of different patients and their conditions before they entered the hospital. If, for example, we're thinking about a person with a heart condition, they may enter in fair condition, serious condition, or critical condition.

Suppose that in Hospital A, most of the patients who enter Hospital A are in fair condition, which is the better condition. So, 700 enter in fair condition; 200 are in serious condition; only 100 enter who start in critical condition. Whereas Hospital B has a different profile of patients upon entry; namely, there are only 100 who enter in the rather benign, fair condition—whereas 700 enter in critical condition.

Now, we could have data about all three of these subpopulations. Look what happens in these subpopulations. For those who enter in fair condition, in Hospital A, 600—that is, 86%—survive; whereas, in Hospital B, those among the 100 who enter in fair condition, 90—or 90%—survive. Of those in serious condition—200 in each hospital—only 100 from Hospital A survive—50% rate—and 150 from Hospital B, for a 75% recovery rate. Then for those who enter in critical condition, only 100 enter in Hospital A, but only 10 of them survive. That is to say only 10% of the critical condition patients for Hospital A survive, whereas, 43% from Hospital B survive.

The overall effect is what we had before. We didn't change the absolute numbers, but by looking at these subcategories, we see that, in fact, Hospital B actually has a higher success rate on each of the subpopulations. This suggests that Hospital B is actually the hospital that you should consider in this serious condition. By the way, this might be the case if Hospital B is, say, a teaching hospital or a research hospital that gets the more serious cases.

So, this is an example that we actually have seen before, and that is Simpson's Paradox. Remember, we saw it before in Lecture 13 when we saw the gender discrimination case and how the view of that gender discrimination example switched when we broke the case down into the subcategories.

Let's move on to another question that we need to answer when we're trying to make health decisions. That's a question that arises when we're trying to distinguish between a quack medicine and a real medicine. There's a comic line that says, "If you take this cold medicine, you'll be well in seven days. If you don't take it, the cold will linger for a week." This is a possible outcome of taking a medicine.

But to determine whether a medicine works, we should compare it to no medicine at all, or to a placebo, as we've discussed before in double-blind experiments. But it turns out that there's a good reason why quack medicines work. Namely, people usually recover from minor diseases without any intervention, without any medicine. So, if you're ill, and you take a medicine that has no effect whatsoever, it's likely that you'll get better.

If you interpret this happy result as an effect of the medicine, what will you do? You'll tell your friends and your neighbors, and soon, a new medical miracle will be upon us. In fact, of course, the medicine may have had no effect at all, or may even have impeded the healing. So, quack medicines appear to work because of a phenomenon called *regression to the mean*.

Let's think about this. If you have a baseline health profile—your usual level of health—if you get ill, then you're going to expect later to return to your baseline, to return to your average. Suppose that we perform a statistical analysis of a totally bogus medicine, and see what happens. The treatment consists of giving the medicine to a person who has, say, a cold, and then seeing whether the person is well in a week. Suppose we try that.

The result will be that we'll get a statistically significant result that more people are well after the week than they were when they started. But it would be wrong to interpret this as an effect of the medicine. Here's a chart that might indicate what I'm talking about.

Let's suppose that these dots represent a sort of random scatter of how you feel from day to day, and the baseline is this horizontal line in the middle, and you sort of randomly feel better or worse from day to day. Suppose on the days that you feel particularly bad—such as this day—you take the medicine. Then look what happens just by randomness. You see all of the six dots—which are the subsequent dots in the week—are higher than where you started. This would happen if we look at a lot of these dots here where you're on the very sick side; you tend to be better afterwards. This is called regression to the mean.

By the way, there's another example of regression to the mean that's sort of interesting, having to do with child rearing. Some people who have reared children sometimes have the impression that what we say to our children has no effect whatsoever on their behavior. Let's just assume for a moment that this is true and look at a statistical analysis that might lead us, actually, to a different conclusion.

Suppose that every time the child does something especially good, relative to that personal behavior, we praise the child, which is a perfectly natural thing to do. Suppose that when the child does something that's particularly bad, we punish the child. Once again, let's assume that neither praise nor punishment has any effect whatsoever on the child.

Let's see what happens. The child will tend to return to his or her average behavior. That's what it means; that's the definition of the average behavior. If something deviates to below its average, it will tend to come back to the average, and if it's above, it tends to go down. So, when we interpret this effect of praise and punishment, let's think it through. If the child is better than average, and we give the child praise, what happens? The next thing that happens is the child returns to average, that is, gets worse. On the other hand, if we punish the child who is behaving worse than normal, what happens? Just by no effect at all, the child tends to return to his or her average and, therefore, it gives the impression that punishment works.

So, it's similar to quack medicine. In both examples, something appears to have an effect, but the actual explanation is regression to the mean. By the way, this occurs in all sorts of areas. Once you get in tune to this concept of regression to the mean, you see it everywhere.

Athletes who appear on the cover of *Sports Illustrated* tend to do worse right afterwards. Why? Why are they on *Sports Illustrated*? Because they did something especially good. What's going to happen? They're going to return to their more average behavior, which means to get worse. There is, in fact, a well-known curse of being on the cover of *Sports Illustrated*.

If you install a camera at a dangerous intersection where there's been a spate of accidents, what happens? Fewer accidents will happen. Why? Because it will return to its average. If you think of it as the opposite, if you put a camera at an intersection that has never had an accident before, then it's sure to have the effect of causing an accident. If ever an accident happens, you can blame it on the camera. You see? So, these are all examples of regressing to the mean.

Another example is the children. Children of tall people, on average, tend to be shorter than those people themselves; likewise, children of short people tend to be taller than the parents. But, they're regressing toward the mean of the overall population. I think this regression to the mean is really interesting.

Let's now turn to one of the most common issues in the realm of personal health management, and that concerns dieting and weight. Many of us have dealt with the problem of keeping our weight at the correct level, and I'm going to use me as an example.

It used to be that I would play racquet sports very vigorously four or five times a week. We'd play squash, and racquetball, and tennis, and so on. My opponents and I would play so vigorously that we had the rule that a victory wouldn't count unless you could make it to the car under your own steam. That was just one of the rules of the game. So, you always had a chance, even if you lost; it sort of gave you hope.

What happened is that I overdid these racquet sports, and my knees gave out, and so I was unable to play racquet sports. But I discovered that I was still able to eat at the same rate as before and, in fact, my eating habits are illustrated by this sweatshirt that some friends of mine gave me that gives an indication of my eating habits. I don't know if you can read it, but it says, "Life is uncertain, so eat desserts first." From my point of view, eating desserts first did not necessarily mean you wouldn't also eat them after the meal, you understand.

In any case, after I quit playing these racquet sports, I discovered that after a certain amount of time, I was about 15 pounds heavier than I had traditionally been. So, every morning, I would weigh myself and keep a record of my weight on a weight chart.

The reason I want to bring this up is the question of how much you weigh sounds like such a simple question, but as a matter of fact, it's not a simple question. What does it mean that you weigh a certain amount? During every individual day, you "gain" and "lose" several pounds; the range is several pounds for every day. One way to take a weight is to weigh yourself in the morning. So what I did is weigh myself before breakfast every day that I was home and that I'd remember, I would put a dot on this chart. The dots on the chart would indicate the reality of my weight. It's not just a single number; it's actually a distribution of numbers.

So, during a period when I really weighed the same, when I wasn't gaining or losing weight, the kind of chart that we would see is a chart like this. We would see a scatter of points—they won't all be the same—that somewhat cluster around a mean value. So, the best description of one's weight would be the mean of those collections of weights. By the way, this is called a *time series*, meaning that at every day, or every time, you take a value—and that's a time series. You can see in days that there's a gap, you would just have a gap and no dots, and then there would be other dots when you get back to weighing yourself.

By the way, this is a histogram of the weight measurements during a period of time, and you can see this is what you'd expect, that the weights would be hovering, having probably some sort of a somewhat normal distribution centered around a single value.

How could you detect whether or not you're losing weight? The way that you could detect whether or not you're losing weight would be to take a time series where every day you take your weight in the morning. If you found that there was a correlation that is statistically different from zero between the days and the weight—and if the least squares regression line has negative slope—then that would be an indication that you are, indeed, losing weight.

A lot of weight loss programs, by the way, tell you only to measure once a week, but I think the reason for that is that they think they're not working with people who are sophisticated in interpreting statistics. Because people who understand that there's going to be variation and that we're just looking at this cloud of points and summarizing them by regression line would actually get better data by taking more weights or time.

It's been my pleasure to talk to you about some issues about personal health. Next time, we will take on the topic of economics.

Lecture Twenty-Two
Economics—"One" Way to Find Fraud

Scope:

Economics is one of the most common arenas in which statistical data are important. Data arise when we measure incomes and wealth, the balance of trade, the deficit, the stock market, the consumer price index, and employment levels. All these data are indications of the economic condition of the world and make real differences in our daily lives. The consumer price index influences such things as tax rates and Social Security benefits. Looking at historical trends is a suggestive method of getting a sense of how the world is changing. However, in some cases, looking at historical trends is a poor method for predicting the future. For example, it is a bad strategy to buy mutual funds based on last year's performance.

A surprising feature of tables of data in economics is that the leading digits of numbers do not occur with equal frequency. This unexpected reality, whose full formation is called *Benford's Law*, gives us an unexpected statistical method for detecting fraud.

Outline

I. Economic indicators tell us something about our financial well-being.

- **A.** The consumer price index (CPI) shows how the value of the dollar changes.
- **B.** The national debt reflects one aspect of the country's financial situation.
- **C.** The Dow Jones Average reflects the business sector of the economy.
- **D.** We will look at trends of these indicators over time to get a historical perspective.

II. The CPI is an important economic indicator.

- **A.** The CPI measures the change over time in the price of the items that people buy in day-to-day living.
  - **1.** The basic idea of the CPI is that it looks at how much various items cost from month to month.

2. If the cost of items goes up, then the CPI records the amount more that we must pay to purchase the same items.

B. The CPI has the goal of giving a sense about how much, more or less, it costs real people in the United States to buy a cross-section of goods and services.

1. To produce the CPI, a specified collection of goods and services is used, called the *market basket.*
2. Each month, people go out and see how much it would cost to buy that market basket.
3. The cost from month to month is compared.

C. There are certainly many features that complicate the simple basic idea of the CPI.

1. One change is that people buy different things as different products come on the market.
2. It would not be sensible to compare only the items in the 1965 market basket with the costs of those same items today.
3. Some items produced then would not be purchased today at all.
4. Many items, such as computer-based goods, would not have been invented in 1965, yet form a significant part of consumer purchasing today.
5. All these features lead to complicated strategies for adding and removing items from the market basket.

D. If, over a certain number of years, it costs twice as much to buy the same collection of things, then we would be able to assert that inflation has doubled the prices. The CPI, then, gives us a sense of what a dollar is worth.

E. When we say that, in 1970, a beginning teacher earned a particular amount and, today, a beginning teacher earns a different, undoubtedly higher, dollar amount, how do we know whether teachers' pay has increased or decreased?

1. We would not know until we compared the value of a 1970 dollar to today's dollar.
2. We would find that the CPI tells us that $1.00 in 1970 is equivalent, in some sense, to about $4.50 today.

3. Thus we could find out if teachers' beginning salaries have, on average, decreased, increased, or remained about the same now as then.

III. Many social programs are legally influenced by the CPI.

- **A.** Social Security benefits for 50 million recipients are adjusted by a formula that uses the CPI.
- **B.** Federal civil service and military pension payments change based on the CPI.
- **C.** The Food Stamp program changes the payment to the more than 20 million Food Stamp recipients.
- **D.** The CPI changes the cost of school lunches.
- **E.** The CPI is used in the federal income tax to adjust tax brackets and the standard deduction.
- **F.** Many collective bargaining agreements involve the CPI to protect salaries and benefits from the effects of inflation.

IV. If we want to understand how various parts of the economy are changing independent of inflation, we can use the CPI to put measures of wealth, income, and so forth in terms of constant dollars, thus allowing us to compare economic conditions over time in a more meaningful way.

V. A graph of the CPI over time tells us the value of a dollar at each time during the last 100 years.

VI. The national debt also shows trends that we can graph over time.

VII. Now we can look at the Dow Jones average, which is the sum of the costs of a particular set of stocks.

- **A.** It shows a consistent increase, and even if we adjust for inflation, we still see a very sharp increase in the late 1990s.
- **B.** Looking at the Dow Jones average causes us to think about stocks and investing.
- **C.** Investing involves looking at data and trying to predict future performance.

VIII. *Data mining* refers to the process of looking at an existing collection of data to find patterns or trends.

A. Data mining can be a very valuable strategy for identifying features of the world; however, there are dangers.

1. In large sets of data, we expect there to be patterns that occur by random chance alone.

2. We would also expect rare events to occur by chance.

B. The appropriate use of data mining is to find patterns, then undertake new experiments to confirm or reject the hypothesis suggested by the mined data.

IX. Suppose we use a data-mining technique as a means to choose a good investment opportunity.

A. Our goal in choosing a mutual fund is to find one that is likely to go up.

B. It is a natural strategy to simply look at which mutual fund increased in value most during the last year and buy it.

C. Unfortunately, that reasonable-sounding strategy is a poor investment strategy.

D. If we look at how well we would have done had we adopted the "buy the best of last year" strategy, we would see that our investments would actually lose money in many years and definitely be a poor investment overall.

E. An analogy that makes this point very clearly is the lottery.

1. Suppose that, among investment strategies, we included buying lottery tickets.

2. We would find that, among the possible investments, there was one $1 investment that earned $100 million.

3. Following the logic of investing in the manner that worked best last year, we would invest in lottery tickets—perhaps we would choose to select the same numbers as the winning ticket.

4. This investment decision is unlikely to be a good one.

X. Data mining leads us to an interesting phenomenon in economics: *Benford's Law*.

A. Physicist Frank Benford did a study of some 20,000 various data sets of numbers and discovered that approximately 30% of the numbers began with 1, rather than about 11%, as expected (1 out of 9).

1. He formulated Benford's Law: The proportion of numbers beginning with 1 is $\log_{10}(1+\frac{1}{1}) = .301$ (around 30%); with 2 it is $\log_{10}(1+\frac{1}{2}) = .176$ (17.6%); with 3 it is $\log_{10}(1+\frac{1}{3}) = .125$ (12.5%); and so forth, until with 9, when the proportion is $\log_{10}(1+\frac{1}{9}) = .046$ (4.6%).

2. If you doubt the veracity of this law, pick a random list of numbers and you will likely see that the number 1 appears disproportionately often as the lead number.

B. Let's take an example. Suppose you deposit $1.00 in a bank account that offers a 10%-per-year growth rate.

1. In looking at your growing deposit through the years, you will find a preponderance of leading 1s at first, because when we start with a number that begins with 1, 10% more still has a leading 1, 10% more still does, and so forth. When you arrive at numbers starting with the digit 2, you start to make bigger jumps, so you have fewer numbers with leading 2s, and so forth until you arrive at the teens; at that point, you are back to leading 1s for quite some time as compared to the 20s, 30s, and so forth.

2. If we begin with a number that starts with, say, 9, then 10% often pushes it beyond starting with 9 and definitely does so on the next occurrence.

C. People who might be called "forensic accountants" have used Benford's Law to detect fraud. When data are false, people tend to make up numbers that have many more leading 5s and 6s than would be expected under the distribution predicted by Benford's Law.

D. Benford's Law is another example in which we can expect regularity in the aggregate that arises from randomness.

XI. This lecture has presented some familiar economic indicators.

A. Much of the way we measure our financial situation involves statistical presentations of the economic conditions of our lives.

B. We must interpret data appropriately to know where we stand. Because dollars change in value over time, comparing a dollar from one time with a dollar from another does not capture the meaning we seek.

C. We must be careful to avoid the pitfalls of drawing inappropriate conclusions from data-mining methods.

Readings:

B. Bowerman, R. O'Connell, and A. Koehler, *Forecasting, Time Series, and Regression: An Applied Approach*, 4th ed.

John A. Paulos, *A Mathematician Plays the Stock Market*.

Questions to Consider:

1. There is considerable debate about whether having a large national debt is bad for the economic health of the country. What data would you gather and what statistical analysis would you undertake to inform your decision on this question?

2. One of the criteria used in stock management literature concerns modifying the risk of a portfolio by including a balance of stable and volatile stocks. What statistical indicators would you look for in data about a stock to assess where that stock fits on the stable-volatile spectrum?

Lecture Twenty-Two—Transcript
Economics—"One" Way to Find Fraud

Welcome back to *Meaning from Data: Statistics Made Clear*. In today's lecture, we're going to be talking about economics. Economics is certainly one of the most common arenas in which we think of statistical data as being centrally important, and it is. We measure incomes and wealth; we measure the balance of trade; the deficit; the stock market; Consumer Price Index; employment levels. All of these data are indications about the economic condition of the world, and these data make a real difference in our daily lives.

We're going to start this lecture, then, by talking about the Consumer Price Index—how it's defined and how it's used, what it means. The Consumer Price Index shows how the value of the dollar changes over time. It's abbreviated CPI, Consumer Price Index, and it's a very important economic indicator for many reasons that I'll explain in a minute.

The Consumer Price Index makes us deal with the challenge of measuring how the price of items alters from year-to-year, or month-to-month, when people buy things for their daily living. The basic idea of the Consumer Price Index is that it just looks at how much various items cost from month-to-month, and compares them. If the cost of items goes up, then the Consumer Price Index records the amount more that we must pay to purchase the same items. That's the basic idea. Its goal, anyway, is to give a sense about how much more or less it costs real people in the United States to buy a cross-section of goods and services. In other words, it's just a measure of the cost of living, over time.

To produce the Consumer Price Index, there's a specific collection of goods and services that's called the *market basket*. This is going to set the standard. Each month, employees of the Bureau of Labor and Statistics go out and see how much it would cost to buy that market basket of goods. They go to different cities in the country and so on, and they find out how much it would cost to actually go into the stores and purchase those exact items. Then the cost from month-to-month is compared. That's the basic idea of the Consumer Price Index, and it sounds like a simple idea. But in reality, there are a couple of wrinkles that complicate the situation, as you might imagine.

One is that people buy different products over time because different products are on the market. It wouldn't be sensible to compare only the items in the 1965 market basket to the costs of those same items today. For example, some items that were common purchases in 1965 would barely be available today. This orange drink Tang comes to mind as a possibility. I think that was probably around in 1965. But many items today, on the other hand, such as computer-based goods, weren't even invented in 1965, and yet now, of course, they form a significant part of consumers' purchasing.

All of these features lead to complicated strategies for adding and removing items from the market basket—but, nevertheless, the concept is the same. The concept is that if, over a certain number of years, it costs twice as much to buy what you might call morally the same collection of things, then we'd be able to assert that inflation has doubled the prices. So, the Consumer Price Index is the thing that measures the value of the dollar. It tells us what a dollar is worth; it's measuring inflation over time.

Let's say that we wanted to compare how well compensated people are in different sectors. We might ask ourselves, "Well, how much was a beginning teacher in some city paid in the year 1970?" Then, we ask how much is that same beginning teacher paid today. Of course, the actual dollar amount today will be a much higher number. But the question is, "How can we make a comparison that would be meaningful to know whether the person is actually being paid more or less relative to the ability to purchase things?" In order to do that, we would have to look at the Consumer Price Index.

Here's a chart of the Consumer Price Index, and if we looked at this chart, this is based on dollars that were viewed as the amount that the actual dollar was in 1983. In 1983, it's standardized at 100. We could choose any year to make as the baseline year for the Consumer Price Index because everything is relative to some arbitrarily chosen year. In this graph, it's the 1982–1984 timeframe that's viewed, and in 1983 we sit at 100. That's the base of this Consumer Price Index.

Then, if we wanted to see, for example, how much a dollar in 1970 was worth compared to a dollar in the year 2000, we would simply go up from 1970, see what the Index indicates, and then go up to the year 2000, see what the Index indicates, and see what the ratio of the actual dollar change is from one time to the other. So, between 1970 and the year 2000, there's roughly a ratio, given by the Consumer Price Index, of about 4.5:1. In other words, $1.00 in 1970 had the purchasing power of about $4.50 today.

So, that is the kind of equivalency that you would want to use if you wanted to argue that teachers' beginning salaries today were, on average, either more or less than they were before. Is the teaching profession valued more or less than it was in 1970? That's the kind of comparison that you would do.

The Consumer Price Index, by the way, is used in many arenas. Many social programs are legally influenced by the Consumer Price Index. All 50 million people who get Social Security benefits, the amount they get is adjusted by a formula that uses the CPI as one of its features. Federal, civil service, and military pension payments change, based on the CPI. The Food Stamp program changes the payment of more than 20 million Food Stamp recipients, based on the CPI. Changes in the cost of school lunches are influenced by the CPI. It's used in the Federal Income tax to adjust tax brackets and the standard deductions. These are all parts of the law that specifically refer to the CPI. In a lot of collective bargaining agreements, they will involve the CPI to protect the salaries and benefits from the effects of inflation.

If we want to understand how different parts of the economy are changing independent of inflation, we would use the CPI to put the measures of the wealth and the income, everything, in terms of constant dollars. Then we can compare conditions over time in a more meaningful way.

One thing that this does is, when we look back nostalgically at old times and we say, "Oh, I could have bought a cup of coffee for 25 cents back in 1903," it gives you a more refined sense of how to evaluate the real buying power of a dollar at different times.

One thing that we can look at is other trends in the economy. The national debt is an example of a trend that we can graph over time. This is a graph of the national debt over the century. But if we wanted to see this national debt from the point-of-view of constant dollars, we could adjust the national debt using the Consumer Price Index to get a different graph. Here is the different graph, adjusted for constant dollars. This is the national debt in terms of 1983 dollars.

You can see some interesting things just by looking. The national debt was rather low until this point. In 1940, you see there's a very sharp spike, which is understandable. That comes from World War II; there was a big spike in the national debt. Notice that from the period of about 1950, all the way up until 1980, notice that the national debt was really almost stable, if we put it in standard 1983 dollars. That's rather interesting that we have this long period of stability, and then in 1980, it starts to shoot way up. This is an example of a way to look at a trend and view it from a different perspective.

Here, we can look at other kinds of graphs, such as the Dow Jones average. You see that this is the Dow Jones average in the dollars of the day; it shows a consistent increase. If we put it in constant dollars, adjusting for inflation, we'll see that even after the adjustment, we still see a very sharp increase in the Dow Jones average during the late 1990s, which was this big bubble in the stock market.

But, let's see if we can bring this discussion down to more immediate issues, and that is for an individual investor. An individual investor in the stock market really wants to choose investments that will go up in short periods of time—so looking at a century may not be of most pertinence. What you want to do if you're trying to invest is you want to look at the data to try to predict the future performance of things.

A very natural strategy, of course, is to look at past data for guidance about the future. What else are you going to look at, in a sense? But when you do that, you should do it with caution. *Data mining* is the term that refers to the process of taking an existing collection of data, and looking at it to find patterns or trends. Data mining can be a very valuable strategy for identifying features of the world, but there really are dangers, and so we want to talk about a few here.

The main danger is, if you have a large dataset, then you should expect there to be patterns that occur by random chance alone. So, if you're looking for patterns, you'll probably find them. Another thing that you can expect if you have very large datasets is that rare events will occur just by chance alone.

The appropriate use of data mining is to find patterns and then undertake new experiments to confirm or reject the hypothesis that's suggested by what you find from this data mining. But it's tempting to celebrate too soon or to take action too soon.

For example, suppose that we want a data mining technique to choose good investments, and our goal was to choose a stock that was likely to go up. Here's a very natural strategy. We could simply look at which stock increased in value during the last year, and buy it. It's a simple strategy. Unfortunately, although it's a very reasonable-sounding strategy, it's a very poor investment strategy. You would find that you would not, in fact, have done well to use the 'buy the best of last year' strategy.

An analogy makes this point even more clearly, and that is to think about the lottery. Suppose we didn't restrict our investment options just to the stock market, but instead, we also included all sorts of different possible investments. We would find that among all the investments that anybody ever did last year, the best investment was to pay $1 for the winning lottery ticket and earn $100 million. That is a really good investment.

Following the logic of investing in the manner that worked best last year, we would just buy a lottery ticket, maybe even buying the exact lottery ticket with the same numbers that won last year, and feel that we were on our way to riches. By the way, it wouldn't be any worse to buy exactly the same numbers that occurred last year; they would be just as likely to win and, of course, just as likely to lose. Obviously, buying the lottery ticket is generally not a good investment.

So, you might prefer, instead of picking an individual stock, to try to predict the whole market trends, and then use the image of the market trends to make a decision. One way to do this would be to look for some other trends that matched the rises and falls of the market. If we could find something that exactly predicted the way the market works, then we could use that as a predictive tool, and follow its pattern.

It turns out that people have actually done this, and looked for the best predictors, and they found that there was a very good predictor for the Standard & Poor 500 Index. That was the butter production in Bangladesh. It turned out to follow the Standard & Poor 500 just exactly right. Why? Because if you mine data, a lot of data, you will find correspondences that are just ridiculous, but they happen accidentally.

A good use of this data mining, by the way—which you probably have heard of—was involved in leading to the best-selling book, *The Bible Code*. That was an example of data mining. You had all this data, and you just searched in it, and, sure enough, you found all of these different patterns. At least it led to a best-selling book, so maybe it's not so dumb after all. In any case, data mining has its dangers.

Now we come to the good part of the lecture. Over the course of these 24 lectures on statistics, there have been some examples of unexpected phenomena—but I think the record is this one right here that I'm going to be talking about today. That is the issue of Benford's Law.

When we think of economics, the first thing that we think of—when we strip away all of the extraneous issues, such as the meaning of the numbers—is that economics has a lot of numbers, long lists of numbers. That's somehow fundamental. Our whole course has been about making sense of the meaning of the numbers. But for the moment, instead of looking for the meaning, let's just look at the numbers themselves. That is, we're going to look at the digits that make up the numbers on these lists. This sounds like a completely crazy thing to do; but I want to tell the story of how this surprising observation—which I haven't told you yet—came up.

It came up in 1881. There was an American astronomer by the name of Simon Newcomb, who noticed that the logarithm books that they had in those days—they had tables of logarithms. Remember, of course, back in those days, people had to do computations using logarithms and there were no calculators or computers, so in order to multiply numbers, or to raise numbers to a power, you would look up the number in the logarithm table. The logarithm tables would allow you to add numbers instead of multiplying them, and then you'd look up the antilogarithm to do a multiplication.

Well, Simon Newcomb noticed that the logarithm tables that were lying around had the property that the front pages of the logarithm tables were more soiled, more worn, than the later pages of this logarithm book. This is very puzzling. It means that the numbers that started with the numbers like 1 and 2 were more frequently looked up than numbers that started with 8s or 9s. He wrote a paper about this observation, and the paper was ignored in 1881. Then 50-some years later, in 1938, this same peculiar observation was repeated by Frank Benford, the one for whom the law was named—and I still haven't told you what the law is.

In 1938, Frank Benford did a study of some 20,000 datasets of numbers. The data came from a variety of sources, economic issues to sizes of lakes, and all sorts of crazy random things, numbers that appeared on the first pages of newspapers, all sorts of random things. He discovered that in most of these cases, approximately 30% of the numbers begin with the digit 1, rather than being evenly distributed. You would think that you should expect about 11% of the leading digits of every number to be 1, 2, 3, 4, 5, 6, 7, 8, or 9. There are nine possible numbers, and so you'd expect 11%, if these numbers were evenly distributed over starting values from 1 to 9. But, in fact, that didn't happen.

He made graphs of these and calculated the percentage of leading 1s, leading 2s, leading 3s, and he found out that, in fact, instead of being an even distribution of this form, instead, he discovered that in his long list of numbers, the proportion of the numbers that start with 1—in almost all of these sets of data—turned out to appear 30.1% of the time. The number 2, as the leading digit of these lists of numbers, appeared about 17.6% of the time. A leading 3 appeared 12.5% of the time. You can see that these numbers are decreasing. The percentage of time that the leading digits are these various digits decreases down to the number 9 appearing only 4.6% of the time.

This law is actually made into a formula here where the fraction of the numbers in a list, starting with the various digits, the digit d, is the log base 10 of 1 plus 1 over d. For example, for the number 1, the log base 10 of 1 plus 1 over 1—that's the log of 2—is 30.1%. Likewise, 1 plus 1 over 2, which is 1-1/2, the log base 10 of 1-1/2 is .176, 17.6%, and so on. This is the actual mathematical description of the fraction of the digits that start with these different digits, and here are the percentages actually listed.

This seems crazy, and I would think that many of you who are skeptical would say, "You've lost your mind. Why in the world would I think that just random data would have this property?" So, the first thing I want you to do is to invite you personally to just do it. In other words, go to some thing, some book, like an almanac or go to the Web, and look for lists of numbers, more or less randomly—it's not true, by the way, that every single set of numbers has this property. But, if you take most economic numbers, or even the areas of lakes in the country, or the populations of counties in the country, more or less anything you can think of, and just look, and you will see that indeed the number 1 appears a much greater percentage of the time than you would expect; it appears about 30% of the time.

Here is an example, just to show you. This is a list of the gross domestic product in the year 2004. It just lists all the countries in order of their gross domestic products. Here, we've highlighted those countries whose gross domestic product starts with a 1. You can see that on this page, quite a few of them start with a 1. Now go to the next page. These start with a 1. Look at this next page, these all start with a 1. Go to the next page. The next page doesn't have any, but on the next page, almost all of them start with a 1. Of course, these are in descending order of gross national product, so you'd expect the 1s to be clustered.

The point is that in that list, you saw a lot of highlighted numbers and, in fact, we've made a histogram here of the percentage of these gross domestic products that started with the different digits—1, 2, 3, 4, 5, 6, 7, 8, and 9. You can see that the histogram—that is, the bars—are telling the actual values from that chart, and then this curvy line here is telling the values expected from Benford's Law. You can see that, in fact, the fraction of the leading digits does seem to accord to Benford's Law. It's rather interesting.

This is the same kind of chart, looking at the population of world countries. It's the same kind of chart, and it shows what fraction of those populations start with the digit 1. This seems so crazy, but this is true. You can see that these populations follow pretty closely the prediction from Benford's Law.

Of course you must be asking, "Why is it possible that Benford's Law is real?" Why is it possible? I'm going to give you an indication of why you might expect more 1s than you think. Here is the way I want to think about it. Suppose that you have a bank account, and you put in a certain amount of money, and every year it grows by a fixed percentage. I know bank accounts don't usually make 10%, but let's just assume 10% because it's easy to compute. Suppose that every year, you put in some money, and then the next year, it's 10% more. Let's just see what happens to the leading digits.

Here we go. Suppose you put in $1.00 as your initial investment in this bank account. Then these numbers here will show you the amount of money in your bank account in the various years, reading across like this. You start with $1.00, go to $1.10, $1.21, and so on. Every year, it's increasing by 10%.

For the first eight years, you always still lead with the digit 1. But look what happens when you get to the digit 2. When you have $2.00—and this is $2.14—in your bank account, and you get 10% more every year, notice that you make bigger jumps than you did when you only had $1.00. From $1.00 to 10% more was a jump of only 10 cents, whereas from $2.00, 10% of that jumps you to more than 20 cents.

Notice you only stay in the 2s then for four numbers before you leap into the 3s. Then every 10% increase increases the 3s by at least 30 cents, so it stays in the 3s for only three 3s before it gets to the 4s. The 4s then go up by 10%, leaping into the middle 4s and then into the 5s; the middle 5s and then into the 6s; the middle 6s into the 7s; and then the 7s—you see, 10% of $7.00-something is at least 70 cents, so it goes all the way into the 8s; there are two in the 8s; then it goes into the 9s immediately, and then we're back to the 1s. Look what happens when you've now accumulated $10.00. Ten percent of that only adds a 1 in the second digit, so it stays in the 1s for much longer before it gets to the 2s. You see?

So, this is an example that shows why it is that if you have exponential growth, at least, you can see why it is that you would expect Benford's Law to accrue. Namely, the bigger numbers, the ones that start with 8, are going to come up much less often because you leap over into the next digit much quicker.

So, this is an example—and you can see I highlighted the values as we go on—that once again, 30% of them will start with the digit 1. Here, once again, is the histogram of actual values in that chart with their different proportions of leading digits, and then the prediction from Benford's Law, which you see it follows quite closely.

There's something interesting about Benford's Law that you might wonder about. I hope that you're at least somewhat skeptical still about Benford's Law because it seems so crazy. But one thing that you might be skeptical about is, it seems as though Benford's Law couldn't be right because suppose you change from dollars into some other currency, for example. That same bank account, you might think it's just going to be off; it would no longer follow Benford's Law. But look what happens if we, for example, consider changing each dollar to a rupee. You see? This exchange rate, by the way, is not even true today; I used it only because 40 was easier to multiply in looking at it than the actual exchange rate.

Look what happens if we take our chart of values, the same values that we had in our bank account in dollars, and we changed them to rupees. Look what happens. What happens is that when we get to $2.50, when we multiply by 4, we get something that starts with a leading 1. Then all the way down to 2 times as much—that is, from $2.50 down to $5.00, all of those multiplied by 4 will lead to having a leading digit of 1. All of the numbers that originally started between $2.50 and $5.00, all when multiplied by 4 give leading digits 1 in the rupees column. It, in fact, keeps the same numbers. That is to say the proportion in the rupees will still be 30% leading with 1s, even though you've changed from dollars to rupees.

It turns out that this weird observation is not only intriguing, but it's also useful. People are using Benford's Law to detect fraud. You see, because when data are made up, rather than coming from reality, people tend to make up numbers that are more evenly divided. In particular, they tend to use a lot of 5s and 6s because they're in the middle. You can compare the number of leading digits in an accounting sheet to Benford's Law. Then any material deviation from Benford's Law is an indication that you better audit that book a little bit more carefully.

So, Benford's Law is another example where we can expect regularity in the aggregate that arises, actually, from randomness. As always, understanding what to expect from randomness allows us to compare expectation with data in order to find convincing inferences—and among them in this case, inferences about possible fraud. I think it's an extremely surprising fact about the world of numbers, which you can personally investigate on your own.

I'll see you next time.

Lecture Twenty-Three
Science—Mendel's Too-Good Peas

Scope:

Advances in empirical science depend on drawing deductions from data. In many cases, a scientific theory is tested by comparing experimental results to predictions of the theory. Randomness can enter in two ways. First, measurement error (noise) in experimental results adds randomness to otherwise definite predictions. Second, some theories, such as Mendel's theory of trait inheritance over generations of pea plants, are inherently probabilistic. In fact, reported results with too little fluctuation can be evidence of fraudulent data. On the other hand, a measurement much different from other measurements (an outlier) can indicate either that some gross error in measuring has occurred, in which case the measurement should be discounted, or that some fundamental assumption is incorrect. Study of the ozone layer in the atmosphere supplies a cautionary example.

Outline

I. Statistics is involved in essentially all scientific matters, from weather reports to quantum physics.

- **A.** During the last 400 years, we humans have fundamentally altered our conception of the universe and our position in it, in many cases as a result of some scientific advance that ultimately is based on the analysis of data.
- **B.** In this lecture, we'll look at several examples of scientific developments and the role of statistics in them.

II. The first example involves Johannes Kepler, the famous astronomer, working in about 1600 as assistant to the astronomer Tycho Brahe, who had amassed vast amounts of data about the locations of planets and stars.

- **A.** Kepler computed that the data fit a model of the solar system in which the planets revolve around the Sun following elliptical orbits.

1. In devising his laws of planetary motion, Kepler used the statistical technique of creating a mathematical model to summarize the data.
2. Later, Isaac Newton formulated his universal law of gravitation, which implies that two masses will follow elliptical orbits about one another.

B. Science frequently progresses in this way.

1. Observations are made that are well summarized by a mathematical equation or model, based on statistical curve-fitting techniques.
2. Later, a more basic understanding of causes and effects can explain that physical model.

III. Hubble's observations about the red shift in spectra from receding stars form another example of statistics playing a prominent role in science. The pattern for a receding star is shifted toward longer wavelengths. The faster the star is receding, the greater the shift.

IV. Another example in astronomy is the 3-degree radiation left over from the Big Bang.

A. Researchers were trying to build a precise radio telescope and kept sensing background noise.

B. After many attempts to fine-tune their instruments to avoid that "error," the researchers discovered that the background noise was a real phenomenon: the 3-degree Kelvin radiation left over from the Big Bang at the creation of the universe.

V. Randomness is at the heart of quantum physics.

A. Modern theories of physics postulate the very unintuitive concept that a subatomic particle, such as an electron, is not in a precise location at a particular time.

B. Instead, the location of an electron is a probability distribution.

C. These theories put the statistical and probabilistic nature of existence in a fundamental position in our understanding of the world.

VI. Measurements and interpretations of measurements are very basic to the scientific process. For example, measurements of the thickness of the ozone layer in the stratosphere or upper atmosphere illustrate another aspect of statistics.

A. When data on this subject were collected by satellite in the 1970s, the values near the South Pole seemed surprisingly small.

B. At first, these were deemed to be bad readings, reflecting some problem in the measuring process, and were omitted from the data summaries.

C. With later measurements, however, it was discovered that the measurements were correctly reporting a real phenomenon, the ozone hole.

D. When one has a lot of data and most of the data are consistent, decisions must be made about what to do with the outliers.

VII. Science proceeds by developing models based on data, then testing the models by comparing experimental results to predictions of the model. In many cases, a scientific theory is tested by comparing experimental results to predictions of the theory.

VIII. Another example of statistics in science is Mendel's famous experiments with peas.

A. Mendel noted statistical patterns in data concerning hereditary traits of pea plants.

1. Yellow is the dominant gene.
2. If homozygous yellow pea plants (those with two yellow genes) and homozygous green pea plants (those with two green genes) are crossed, the first-generation offspring all look yellow, being yellow heterozygous plants (those with one yellow gene and one green gene).
3. The second-generation offspring, however, which are the result of crossing yellow heterozygous plants, are about one-quarter green, indicating homozygous green pea plants.

B. This was the fundamental observation that led to the concepts of genetic inheritance and dominant and recessive genes.

1. Two genes, one from each parent plant, combine to form the genetic makeup governing the color of the offspring.
2. Yellow is the dominant gene; thus, only if both genes in a plant are for green will the plant be green.
3. Assuming that one of the genes is randomly selected from each parent plant, we would expect that *about*, but not exactly, one-quarter of the time, both contributions will be green.
4. To determine whether a yellow plant was heterozygous or homozygous, Mendel took the yellow plants and bred them with themselves 10 times. If the plant was homozygous, on each of those 10 times, he would always get a yellow plant. However, if he had a heterozygous plant, he reasoned that the chances were very good that in 10 breedings, 1 of the self-breedings would contribute both green genes, and the plant would come out green.
5. If we performed many experiments, with 800 yellow plants in the second generation, we would expect different numbers of homozygous yellow plants in different experiments, with the center of the distribution around 200 but with occasional outliers.

C. Statisticians have looked at Mendel's reported results and have discovered that it would be very unusual, given the amount of his data, that all of the results would be in the narrow bounds he reported. In short, Fisher believed that the data that Mendel got were too good.

1. Ronald Fisher, to whom we were introduced in Lecture Twelve, found that Mendel's results lie within 1 standard deviation of the mean much more often than the expected 68% of the time.
2. Fisher also noted that Mendel used the method of cross-breeding yellow plants with themselves 10 times to identify whether they were homozygous or heterozygous.

3. In an interesting twist, this method implies that we should have expected Mendel to misclassify a certain percentage of plants; however, Mendel's reported data are closer to the data expected if all the plants were classified correctly.

D. Mendel's work illustrates all aspects of statistics, including design of experiments and interpretation of data, and may illustrate the possibility that the data were made to look somewhat better than they actually were.

IX. Using carefully executed statistical capture-recapture methods, scientists can estimate quantities as diverse as the population of tigers in a jungle, the volume of water in a lake, and the size of a natural gas deposit in the ground.

X. Statistical analysis of experimental data is key to validating or invalidating a scientific theory.

Readings:

E. T. Jaynes and G. Larry Bretthorst, eds. *Probability Theory: The Logic of Science*.

R. A. Fisher, "Has Mendel's work been rediscovered?" *Annals of Science* (available at

http://www.library.adelaide.edu.au/digitised/fisher/144.pdf).

Questions to Consider:

1. Measurements are never exact. As instruments improve, would you expect the distributions of measurements of physical constants to have less variation, have a different mean, or both?

2. Suppose data are found that reject a scientific theory with a high level of statistical significance. Under what circumstances would you tend to reject the data rather than reject the theory? Is that ever a good idea?

Lecture Twenty-Three—Transcript
Science—Mendel's Too-Good Peas

Welcome back to *Meaning from Data: Statistics Made Clear*. In this lecture, we're going to be talking about science and applications of statistics to science. Certainly, statistics and the statistical analysis of data are obviously central players in all aspects of science. During the last 400 years, we humans have fundamentally altered our conception of the universe and our position in it, in many cases as a result of a scientific advance that ultimately is based on the analysis of data.

Let me first talk about the concept of the solar system. In the year 1600, Johannes Kepler was in the observatory of Tycho Brahe and was looking at the data associated with the positions of the planets as they moved in the night sky. Tycho Brahe had accumulated the best observational data—this was before the invention of the telescope—that had ever been collected. One of the anomalous challenges was to explain the funny behavior of the motion of the planet Mars, which had this retrograde motion and other anomalies that made it difficult to explain.

Kepler took the data from these observations, and basically fitted a curve to the data to make his laws of planetary motion—namely that the planets revolve around the sun in an elliptical form. So, Kepler's contribution to our understanding of the solar system and our place in it was a statistical one. It was data fitting, finding a model that approximated the data that was actually observed.

Later, our understanding of the reason for having elliptical orbits was put on a different foundation when Newton proposed his universal law of gravitation, from which it follows as a mathematical consequence that planets that were going this kind of inverse square law of gravitational attraction would, in fact, automatically follow elliptical orbits. So, that changed the presentation and our concept of the reality from a statistical curve-fitting concept of Kepler to one based on a more fundamental physical law.

This is frequently the way that science proceeds—that at one point, we have an observation that's a rather empirical kind of observation that later is put on a firmer foundation when a theory that implies that foundation is found. In our concept of the entire universe, we have the example of Hubble's red shift, which was a statistical observation that there were differences in the spectra of stars that were further away that implied that stars further away were moving faster.

There was the discovery of the 3-degree background radiation that was discovered when it was found that in trying to create a very precise radio telescope, the people were unable to tune it properly to the level that they thought. No matter what direction they turned it in the night sky, they discovered this background radiation. That was a statistical or a measurement issue, which was then later explained as the leftovers from the big bang. It was confirmation of the big bang theory of the universe.

So, measurements are used in both finding out interesting things about the universe, and also to confirm a model or reject a model. For example, Einstein's Theory of General Relativity was confirmed—or we should say Newton's theory of the universe was rejected—when observations were made about the light bending around the sun, in order to say that the measurements that were obtained would not have been obtained had Newton's theory been correct.

In physics, the concept of the most fundamental understanding of the basic particles that we live with, or are made of, are probabilistic in nature. Quantum field theory implies that we are not to think of a subatomic particle as being in a particular place at a particular time—but, instead, it's a probability distribution, and that that is the actual reality. This very strange idea puts the concepts of probability and chance as fundamental players in our understanding of the physical world.

Measurements and their interpretations are very basic to the scientific process. I want to give an example where measurements had an interesting wrinkle to them. Back in the 1970s, people began measuring the thickness of the ozone layer in the stratosphere, or the upper atmosphere. I want to tell the story as an illustration of some part of a statistical practice.

When measurements were made about the thickness of the ozone layer, there were many measurements that were very similar to each other, as you would expect if you were measuring something where the concept was that the thickness was uniform. Many of the measurements were the same, but there were a few measurements that were much, much smaller—that were close to zero. There was a computer program that was taking these measurements, and it was set to identify measurements that were too far out of the bounds of the preponderance of the evidence, and label them as potential outliers. Outliers are, remember, data that don't seem to fit the pattern of the remaining part of the collected data. So, by noticing that there were these measurements that were close to zero, the computer program threw them out as potential outliers. It actually kept the data, and the data were recalled later, which was important.

The issue is that in collecting data, very often, there are errors that occur accidentally. For example, our measuring instrument may have been on the blink that day, and got a value that was way off. Before we had an understanding that there was an actual hole near the North and South Poles in the ozone layer, the hypothetical image of our concept of the ozone was that you could imagine that there was a uniform distribution of the ozone layer across the whole Earth.

So, one question that we have about data is, "What is the generality of the application of the data that we collect?" In other words, if we collect data from various spots on the Earth, and it was pretty uniform, suppose no data had been collected around the poles at all. Then it might have well been that we would claim that the data implied that there was a uniformly thick layer of ozone across the whole Earth. The fact that there was this anomalous outlying data, which we then threw out, turned out to be actually a real phenomenon, and we shouldn't have thrown it out, and it actually pointed to a phenomenon that is actually real.

I'm trying to make the point that sometimes the question of the applicability of the data we get is not clear, how generally applicable it is. Suppose that we assume that all fish are the same with respect to some property. Then, we might do an experiment that just used, say, a salmon, and then draw conclusions about all fish. That's certainly fine if all fish really are the same with respect to that property; but it's not always clear that that is the case. There could be some lurking variables that make different species of fish behave differently.

So, in any case, science proceeds by developing models of what we think the world is like and then looking for data that test the models by comparing the experimental results with the predictions of the model. In many cases, scientific theory is tested by comparing experimental results to the prediction of such a theory.

One famous example of that is in the case of Mendel's famous experiments with peas. In the middle of the 19th century, Gregor Mendel was a monk in a monastery. He conducted many experiments that are now viewed as seminal experiments with respect to our understanding of heredity, and how characteristics of the parents are passed on to the children.

In particular, Mendel had a model in mind where he envisioned that each of the parents had two genes, of which they contributed one to the daughter plant, and then those two genes made up the genetic basis for the actual appearance of the plant.

So, what I'm going to be doing now is explain some of Mendel's experiments, and try to describe how it is that those experiments proceeded. We'll begin with this chart. Here's an example of a typical kind of experiment that Mendel performed. The peas that he was working with had actually many different characteristics that he worked with, but I will just talk about one characteristic—namely, color. He also talked about things like whether it was a wrinkled pod or a smooth pod, and other kinds of characteristics, but I will just talk about the two different colors, yellow and green.

The basic concept of his experiment was the following. He imagined that he had some plants that had two yellow genes. He took these yellow genes, and then he crossed them with plants that he assumed to have two green genes. It turns out that yellow is the dominant gene so that if you have a plant that has a yellow and a green gene, or has two yellow genes, then the plant will appear yellow.

So, when he crossed plants that had both yellow genes—and this is called, by the way, *homozygous*, meaning that both of the genes are the same—with plants that had both green genes, all of the offspring had one of the genes from this parent and one of the genes from this parent in its genetic makeup. All of the offspring of these plants had one yellow gene and one green gene. All of them appeared yellow because yellow is the dominant gene.

The interesting part of the experiment occurs at the next generation. Suppose that we take plants that were the result of that previous breeding, and we had plants, each of which was *heterozygous*—meaning they had a yellow gene and a green gene—and we combined them together to form the potential offspring that they could have. The theory that Mendel proposed—which is what these experiments were intended to support—would predict that for each parent, it would randomly contribute one or the other genes to the daughter plant.

So, there are four possible things that could happen in this kind of an experiment. First of all, you could have both parents contributing the yellow gene; or the first parent could contribute green and the second yellow; or the first yellow and the second green; or both of them could contribute green genes.

Only in this corner here, the plants that have two green genes, would the plant actually appear green—this is the pod of the pea plant. It would appear green only if both of its genetic contributions from the parents were both green genes.

The way that these experiments proceeded was that Mendel crossed a bunch of heterozygous plants, and looked at the proportions of the offspring plants that were yellow, and the percent that were green, with the expectation that, in fact, 1/4 of them were green, and 3/4 of them were yellow. That was his expectation, and he was trying to prove this theory by getting evidence by doing these crossings. Among these three, he expected 1/3 of these to be homozygous—with yellow genes—and the other 2/3 to be heterozygous—one green and one yellow gene.

Let's look at some data. Suppose we had done an experiment in which there were 200 plants expected in each of these quadrants. Then, we would expect that in doing this experiment many times—which he did—he would have expectations of this kind of an outcome, if this were the size of the experiment.

However, you might ask the question, "How does he know whether a plant is heterozygous or a plant is homozygous?" Mendel specified his method for determining whether the plant was homozygous or heterozygous. His method was that he took these plants, these three that appeared yellow, and he bred them with themselves 10 times. By breeding them with themselves, of course, if it's homozygous, on each of those 10 times, he would always get a yellow plant. However, if he has a heterozygous plant, he reasoned that the chances were very good that if, in 10 breedings, one of the self-breedings would contribute both green genes, and the plant would come out green. That would be an indication that the plant that he started with was a heterozygous plant, that it had the green gene as well as the yellow gene. His method was to take his plants and do 10 self-cross breedings, and on that basis, classify the plant as being heterozygous or homozygous.

This was the data that he collected. He collected a great deal of data, which all supported his theory. In many instances, in data of this size, he would find that there were 201 plants that he had classified as being homozygous with yellow, and that the ratios were appropriate, very close.

In 1936, Ronald Fisher—whom you may remember as the statistician of lady-tasting-tea fame—wrote a paper in which he investigated Mendel's data. He made some observations about the quality of Mendel's data. In particular, he noted that Mendel's data was too good to be true. Remember that when we are dealing with a random process, we don't expect the answers to always be exactly according to expectation. We expect a distribution of the answers. Most of the time, the answers will be within a certain distance of the expectation, but a certain fraction of the time, we would expect to have outliers. We would expect to have unusual and rare occurrences.

One of the things that Ronald Fisher pointed out was that the number of experiments in which Mendel's data was very close to expectation was too great to be believed. Let me give you an example of the reasoning that Fisher used. Suppose that you take a coin, and you flip it 1,000 times. You know that, on average, the mean of the distribution of the flips is going to be 500. But you also know that if you actually flip a coin 1,000 times, often the number of heads will be less than 500, and more than 500. Several times during these lectures, we have seen the distribution of the number of heads, for example, that would be likely to appear if you did the trial of flipping a coin 1,000 times.

This is the distribution. In other words, this tells you that occasionally, you would get 540 heads, but the proportion of times you'd get it would be very rare, and most of the time, you would have some number of heads in your 1,000 flips that lies somewhere in the big part of this curve.

However, this very thin region in the middle here is a region that contains only a small fraction of the data, about 7.5%, and most of the time that you flip a coin, it would not lie between 499 and 501; most of the time you would get values that differed from 500 by some bigger amount.

What Fisher did was to notice that Mendel's data tended to give more outcomes that were within 1 standard deviation of the mean than would be expected. You would expect outcomes of an experiment that involved random chance to lie within 1 standard deviation of the mean—in a normal distribution, about 68% of the time. But that means that about 32% of the time, you would expect the results of that random experiment to have values outside of 1 standard deviation from the mean. Yet it turned out that the data that were reported by Mendel had too high a frequency of good data, of the data being too close to expectation. So, Fisher was arguing that the data was not properly constructed.

But, Fisher went on to make another interesting claim about the results from Mendel's data, which is the following. If you recall, the strategy by which Mendel chose to classify a plant as heterozygous was to do the experiment of self-breeding 10 times. If the plant came out yellow all 10 times, it was classified as homozygous. There is a chance, a small chance, that if one took a heterozygous plant, and by randomness alone, bred it with itself 10 times, every one of those 10 times, it would have contributed a yellow gene, and would be yellow every one of those 10 times.

In fact, it's not a difficult computation to see exactly what proportion of the times that would happen. Namely, when you cross-breed a heterozygous plant—that is a green and a yellow plant, and you randomly pick the two—the chance is 3 out of 4 that 1 of the 2 genes contributed, at least 1 will be yellow. When you cross a heterozygous plant with itself, there's a 3/4 chance that it will be yellow. If you do it 10 times, there's a $3/4^{10}$ chance that it will be yellow every single time, which is 5.6%. The probability of misclassifying a heterozygous plant is 5.6%.

In other words, the method that Mendel used was to take a plant and ask if it's homozygous—having two yellow genes—or heterozygous—having one green gene and one yellow gene. He then bred it with itself 10 times. If it was always yellow, he classified it as homozygous. But 5% of the time that you started with a heterozygous plant, it should have been misclassified as homozygous.

What that means is that Mendel should have—in doing his experiments in the way he specified that he did them—misclassified 22 plants that were, in fact, heterozygous; he should have classified them as homozygous. So, what Fisher pointed out was that the actual expectation for that square of the classification of plants should have wrongly classified 22 extra plants as homozygous that actually were heterozygous. So, the actual expectation from the experiment should have been 222.5, not the 200, which was the actual expected outcome for the plants that are really homozygous.

The effect of this is that when Mendel reported an experiment—and this is a specific example of one experiment of many—in which the reported number, 201, is very close to the quasi-expectation of 200. But you see that it's rather distant from what should have been expected, including the falsely classified heterozygous plants that should have been falsely classified as homozygous using that method.

One can do a statistical analysis where the null hypothesis is that the fraction of the plants should have been 222.5 out of 600; the alternative hypothesis is that the fraction is less than that. You find that the p value for that hypothesis test is .037. If you use the traditional cutoff margin of 5%, you would actually reject the null hypothesis that 222.5 was the actual expected value; but it should have been the expected value.

This is the kind of reasoning that Fisher used in his article to show that the data that Mendel got were too good and, in fact, in this case, Mendel wasn't subtle enough to realize he should have been expecting 222.5 instead of expecting 200 in that box. So, his data came out more like the 200 than what he really should have found.

I want to quote a few lines from this paper by Ronald Fisher. It's titled "Has Mendel's Work Been Rediscovered?" It was published in 1936 in *The Annuls of Science*. I want to read the description of Fisher. He is the Galton Professor of Eugenics at the University College London, by the way. That's another interesting wrinkle to Fisher, which we will not talk about now. What he said in this paper is the following:

> The discrepancy is strongly significant, and so low a value could scarcely occur by chance once in 2,000 trials. There can be no doubt that the data from the later years of the experiment have been biased strongly in the direction of agreement with expectation. [He's saying that the data were cooked.] One natural cause of bias of this kind is the tendency to give the theory the benefit of doubt when objects such as seeds—which may be deformed or discolored by a variety of causes—are being classified. Such an explanation, however, gives no assistance in the case of the tests of gametic ratios [which is what we were just talking about], and of other tests based on the classification of whole plants.
>
> The bias seems to pervade the whole of the data. [This is a quotation from his work.] Although no explanation can be expected to be satisfactory, it remains a possibility, among others, that Mendel was deceived by some assistant who knew too well what was expected. This possibility is supported by independent evidence that the data of most, if not all, of the experiments have been falsified so as to agree closely with Mendel's expectations.

This is a little wrinkle in the history of science associated with data being too good to be true.

I want to close this lecture with one other completely different kind of experiment that's a little on a lighter tone. Here is a basic scientific question that you may have wondered about. How can people count tigers that are in the jungle when they can't find all the tigers? How would you go about counting tigers? There's a very clever way of counting tigers in the jungle, and this is the way you do it.

You capture 50 tigers, and then you put ear tags on those 50 tigers, and then let them back into the jungle. You let them mix around for a while, and then you come back, and you capture some other group of tigers, perhaps 100 tigers. You see what proportion of the newly captured tigers have ear tags. Then the computation is very simple to get an estimate for the number of tigers in the jungle. Here it is.

We've tagged 50 tigers, so you imagine 50 tigers going around the jungle randomly. You randomly capture 100 of them. Suppose, for example, that 8 of the 100 that you recaptured were tagged. Then you would have this little formula here that 50—the number of tagged tigers—is to the whole population as the number recaptured, who were tagged, are to the number that you actually captured. In other words, among the 100 that you captured the second time, 8 of them had ear tags. So, if 8 out of 100 have ear tags, you would expect the whole population to reflect the same proportion. Just doing the math here, we see that the population would be estimated to be 50×100 divided by 8, or 625 tigers in the jungle.

I always found this a very clever method. It's a question of capturing, and then you let them out, and then you capture some more, and you see what proportion of the newly captured ones have the tags. Using the proportions, you just figure the whole proportion is reflective.

You can do the same kind of an experiment finding the volume of water in the lake. You can take 1 pound of salt in this clear-water lake, and dump it into the lake. Wait until it stirs around, maybe several weeks, until it's completely evenly dispersed throughout the lake. Then take 1 cubic foot of water from the lake, and boil off all the water until you just measure the salt. By doing a very, very fine measurement of the salt, suppose that you have 1 millionth of 1 pound of salt in that 1 cubic foot of water. Then you would infer that there were 1 million cubic feet of water in the whole lake.

So, you can imagine doing this capture/recapture method in many instances. For example, if you had a natural gas deposit under the ground, and you want to know how big it was. You could put in a certain volume of some number of molecules of something that mixed around, and if it mixed around, then you could take out a sample from the natural gas, and see what proportion of the molecules were in that volume of that second sample, and deduce what the volume of the whole find was.

So, we've seen, then, in this lecture several examples of statistical applications in science, of which, of course, there are many. This is obviously just the tip of the iceberg in this area.

I look forward next time to talking to you about the use of statistics in discovering the authorship of papers.

Lecture Twenty-Four
Statistics Everywhere

Scope:

Statistics is a subject that permeates essentially every area of our lives and world. It is a powerful tool for seeing our world in a more detailed fashion and for making informed decisions, although its subtleties and potential misuses caution us to avoid thoughtless acceptance of statistical conclusions. The recent and expected future development of computer speed and capacity allow us to imagine using statistics with ever more scope and effect. How much information and understanding can we hope to gather from statistical data? How much more meaning would better statistical techniques allow us to find? Statistics is a tool with wide applicability. It has limits that need to be acknowledged and respected, but its potential for helping us find meaning in our data-driven world is enormous and growing.

Outline

I. Often, data can contribute decisive evidence in an otherwise difficult matter to resolve.

- **A.** For example, during the debate about ratification of the Constitution, Alexander Hamilton, James Madison, and some others anonymously wrote *The Federalist Papers*.
  1. People disputed the authorship of about a dozen of these essays.
  2. Arguments based on philosophy and style were not persuasive.
- **B.** Discriminant analysis, that is, statistical analysis concerning the frequency of the use of specific common words (for example, *on* instead of *upon*, where appropriate) provided powerful arguments for Madison's authorship.

II. This *Federalist Papers* example is satisfying in that it suggests that seeking data to find persuasive arguments is a valuable method for coming to conclusions.

- **A.** Sometimes it is not clear what data are pertinent.

B. In the case of *The Federalist Papers* dispute, it would not be immediately obvious that counting the frequency of trivial words would be the road to decision on the authorship issue.

C. Of course, the technique can be applied to other authorship questions, such as whether the works of Shakespeare were actually written by Marlowe or Bacon and whether Shakespeare wrote a newly discovered poem attributed to him.

1. In the latter instance, the new poem was discovered in 1985.
2. Instead of looking for frequency of common words, as with *The Federalist Papers* controversy, statisticians looked for new, original words because Shakespeare was well known for inventing words.
3. Because 9 new words appeared in this 429-word poem, these statisticians were able to deduce that, in fact, this poem was very possibly written by Shakespeare.

D. The results in those cases, however, do not seem as clear-cut as in *The Federalist Papers* dispute.

E. Data and appropriate interpretation are powerful arguments not easily refuted without further data.

III. The evidence for the determination that Madison wrote the anonymous articles would be evaluated differently by Bayesian statisticians versus frequentist statisticians.

A. A Bayesian would be willing to say that, given the word usage in *The Federalist Papers*, there is a 99.9% chance that it is written by Madison.

B. Frequentists would say that the articles are either written by Madison or not.

C. Both camps would probably agree to a statement something like this: "The probability is only 2.4% of getting no *upon*s when randomly selecting a collection of 1,000 words from a person's writing that generally has 6 *upon*s per 1,000."

IV. Using data and statistical analyses will become an even more prominent part of our world in the future than it is now. The principal reason is the continuing development of computer technology.

A. With the computer, it is now possible to deal with large databases and use techniques that would have been computationally impossible previously.

B. Some such techniques involve simulation as a means to understand a collection of data.

C. Often, such methods are computationally intensive and, consequently, become increasingly valuable as computer power increases.

D. One such technique is called the *Monte Carlo method*, which involves using random processes by a computer to generate thousands of scenarios, enabling statistical techniques to derive the distribution of the behavior of a system.

E. Today on a home computer, it is possible to do amazing statistical analyses essentially instantly.

1. Early textbooks on statistics would emphasize methods for reducing computation.

2. Now, with computers, those techniques are not so important.

V. Let us now consider some observations about the statistical enterprise altogether.

A. First, there is often a difference between statistical knowledge and understanding based on deeper principles.

B. Second, statistics is used more than it is understood, as evidenced in the blind application of statistical tests.

1. Any attempt to reduce data to a formulaic adherence to following tests is likely to be misleading and can often produce nonsensical arguments. Recall the mean levels of wealth of graduates at Lakeside High School in Seattle.

2. Statistical reasoning is subtle and prone to counterintuitive examples; understanding the underlying logic is necessary in order to have confidence in the result. Recall the exercise of choosing the best hospital.

C. Hypothesis testing has issues of its own, such as the arbitrariness of the level of rarity that we deem statistically significant.

D. The persuasive strength of a statistical argument requires a clear understanding of the statistical reasoning, the context of the situation, and the details about the study or data that allow us to interpret the meaning of the statistical data and arguments with conviction.

E. Unless we have high confidence that a survey was conducted appropriately, then the statistical result may not be as strong as reported. Recall the results of the *Literary Digest* poll.

F. Good statistical results and inferences are far superior to anecdotal evidence on an issue, but we need to be critical consumers.

G. Statistical knowledge is, by its very nature, an admission of ignorance. Dealing with statistics often means that we do not have the whole story.

H. Statistics is a collection of profoundly powerful methods for understanding our world with more detail and more meaning.

VI. We have seen statistics as having two basic parts:

A. Organizing, describing, and summarizing a collection of data when we know all the data.

B. Inferring information about the whole population when we have data about only a sample of the population.

VII. When we make an estimate of the value of a feature of the whole population given data about a sample, our challenge is to describe how accurate our estimate is likely to be by answering the following questions:

A. How close is our estimate to the correct value?

B. How confident are we that our estimate is, in fact, that close?

VIII. Statistics is a powerful tool for understanding our world. We end the course with a statistic about the course: Among people who learn something about statistics, 100% appreciate our world with more clarity.

Readings:

Norman L. Johnson, and Samuel Kotz, eds. *Leading Personalities in Statistical Sciences: From the Seventeenth Century to the Present*.

William S. Peters, *Counting for Something: Statistical Principles and Personalities*.

Theodore M. Porter, *The Rise of Statistical Thinking, 1820–1900*.

David Salsburg, *The Lady Tasting Tea: How Statistics Revolutionized Science in the Twentieth Century*.

Questions to Consider:

1. As a practical matter, how can a more sophisticated understanding of statistical reasoning come into play on an everyday basis? Can we apply pressure on the media to make the details of studies more readily available so that we have a better chance of determining whether the statistical reasoning is sound?
2. Being attuned to statistical reasoning adds depth to our appreciation of the world around us. Choose an issue in which you have an interest and find data that are pertinent to your appreciation of it. Then apply the techniques of organizing, describing, and summarizing the data and inferring meaning from the data to understand that part of the world better.

Lecture Twenty-Four—Transcript
Statistics Everywhere

Welcome back to *Meaning from Data: Statistics Made Clear*. Statistics provides conceptual tools with a very wide range of applicability, as we've seen. Often, statistical reasoning can contribute to decisive arguments in matters that seem very difficult to resolve, and even issues that don't appear to have any statistical component to them at all. We'll start this lecture with such an example.

In 1787 and 1788, Alexander Hamilton, James Madison, and John Jay wrote a series of essays about the Constitution, advocating that the Constitution be accepted by the people of New York. These were the *Federalist Papers*. The goal of these papers was to convince people to vote for the Constitution, and they wrote a total of 85 essays. They didn't sign the papers with their names; they were all published under the pseudonym Publius.

But of these 85 essays, the authorship was clearly known in all cases except for about a dozen. Alexander Hamilton was known to have written 51 of these 85 essays; John Jay was known to have written 5; James Madison was known to have written 14; and Madison with Hamilton wrote 3 together. This left 12 disputed *Federalist Papers*. People had disputed the authorship of these essays for many, many years, and some contended that Hamilton was the sole author, and others asserted that Madison was the author. No one thought Jay was—so it was between Hamilton and Madison. Arguments were presented, based on all sorts of things, as you'd expect—namely, what positions were presented in the disputed articles, and issues about their style of writing, and so on. But, none of these arguments was powerful enough to settle the dispute.

In 1964, statisticians Frederick Mosteller and David Wallace published a book called *Inference and Disputed Authorship: The Federalist*, in which they approached the question from a statistical point of view. They recognized that different individuals develop habits of word usage that are rather distinctive from person to person. In the case of Hamilton and Madison, there were many known examples of writing by each of the two. Because they had these sources, counting the frequency of the use of different words was a very possible thing to do.

In particular, Mosteller and Wallace viewed the use of unimportant words as particularly significant, with the idea that the unimportant words were words that more or less just arise independently of the content being conveyed. This method of analysis is called *discriminant analysis*—trying to discriminate between, in this case, the word usage of Hamilton versus the word usage of Madison.

They used many words, but one specific word—to ground our discussion—was the word "upon." Hamilton was fonder of the word "upon" than was Madison. In Hamilton's extant writing, he used "upon" at the rate of about 6 words per 1,000; that is, 6 occurrences of the word "upon" per 1,000 words in his writing, on average—whereas, Madison only used the word "upon" less than 1 time per 1,000 words, overall.

By looking at the actual *Federalist Papers* in question—the ones of disputed authorship—several such words were analyzed, and the relative frequency of the use of those words by Hamilton versus Madison in the previous known writing, versus the unknown writing, that kind of analysis was used to try to distinguish the authors.

The evidence and the reasoning from it are rather persuasive, but I want to demonstrate how this was done by actually looking at the *Federalist Papers*. These two volumes here comprise the *Federalist Papers* and some others. I want to read you some excerpts from some of this writing.

This first excerpt will be from "Federalist Essay No. 23," which is definitely by Alexander Hamilton. On this page, there are three occurrences of the word "upon," and I'll just read some of these sentences. It says, "It rests upon axioms as simple as they are universal, the means ought to be proportion to the end," and so on. Later on the page, he says, "As their requisitions are made constitutionally binding upon the states, who are in fact, under the solemn obligations to furnish the supplies," and so on. Finally, here at the end, a third instance, "If we are in earnest about giving the union energy and duration, we must abandon the vain project of legislating upon the states in their collective capacities."

So, he used the word "upon" several times. If you notice, in the use of this word, in many instances, he might have chosen to just use the word "on." For example, here, "As their requisitions are made constitutionally binding upon the states," you could have said "on the states." But Hamilton used the word "upon."

Let's look at an example that is known to have been written by James Madison. This is "Federalist Essay No. 39." In it, he has the following part of a sentence: "We may define a republic to be—or at least may bestow that name on—a government which derives all its powers directly or indirectly from the great body of the people." You notice that if you chose to use the word "upon" in that sentence, it would sound quite reasonable. "We may define a republic to be—or at least may bestow that name upon—a government which derives," and so on. So, perhaps Hamilton, who preferred "upon," might have used "upon" in that instance. Of course, in any individual case, it's completely inconclusive.

Now we'll turn to one of the disputed essays, which is "Federalist Essay No. 52." This is one of the disputed essays, and here are a couple of sentences in which the word "on" is used, where "upon" might be used instead. "Under the federal system, cannot possibly be dangerous to the requisite dependents of the House of Representatives on their constituents." The author might have said, "requisite dependents of the House of Representatives upon their constituents," which would read rather well.

Here's another example. "The advantage of biennial elections would secure to them every degree of liberty which might depend on a due connection between their representatives and themselves." That might have been written, "every degree of liberty which might depend upon a due connection between their representatives and themselves."

So, we see that in these examples, there are choices being made by the author in each instance of whether to use the word "on" or "upon," and Hamilton more frequently used the word "upon" than Madison did.

In this particular dispute, it was discovered that, in fact, the authorship of these papers—that is to say the word usage of meaningless words—corresponded much more strongly with Madison than with Hamilton. So, this was strong evidence from which it was rather persuasively concluded that, in fact, the disputed *Federalist Papers*—all of them were written by Madison. So, this is an example in which we have an arena in which it's not clear that statistical evidence would be pertinent whatsoever.

Here's another example of analysis of authorship. This is a paper entitled, "Did Shakespeare Write a Newly Discovered Poem?" by two statisticians. You might expect that it would be written by two English professors, but, no, this is written by two statisticians. In this paper, there's a different strategy by which these statisticians try to decide whether or not a poem was written by Shakespeare.

The method that they use in this case is the method of looking for unusual words. The previous strategy was to look for usual words, and how frequently usual words were used. This strategy is another statistical variant in which they look for unusual words.

The kind of statistics that they presented were ones where they tried to claim that, among the words that Shakespeare used, very frequently, he used words very few times. In other words, his vocabulary included 31,534 distinct words that appear in the 884,640 total words in the Shakespearean canon. It's interesting that all of these things are so precise.

The concept here was to look at the words in this newly discovered poem. A poem was discovered by Gary Taylor on November 14, 1985. The question was whether this Taylor poem—as it's come to be known—was or was not written by Shakespeare. It had 429 total words, and the issue was how to distinguish whether or not this could be ruled out as a Shakespeare poem. It's more ruling out than asserting that it was, in fact, written by Shakespeare.

This is the basis on which this analysis is done. They discovered that unusual words are very common for Shakespeare. Two-thirds of the 31,534 distinct words occur three or fewer times in the entire Shakespearean canon. The effect of this is that you expect, even in relatively small samples of works from Shakespeare, you expect words that do not appear in the rest of the canon.

So, in this poem of 429 words, there is an expectation—and these statisticians developed a model—of how many words that they would expect to be new—words that did not appear elsewhere in the canon. In fact, there were 9 new words that appeared in this poem, which are the following: admirations; besoughts; exiles; inflection;

joying; scanty; speck; tormenter; and explain. Those words occurred nowhere else in the canon, which is sort of surprising.

By the way, a variation on a word counts as a different word in this counting. For example, "admirations" counts, although the word "admiration" appears 14 times in the rest of the canon. And "besoughts" counts as a new word, although "besoughted" did appear elsewhere in the canon. So, using this strategy and a data analysis, these statisticians were able to educe evidence that, in fact, this poem was certainly possibly written by Shakespeare.

One thing that is sort of troubling about this—I'm sure to a lot of people who are in the subjects of literature or history—is that it seems somehow wrong that decisive evidence in favor of one theory or another would not be based on expertise in the area in which this evidence is being educed. I wonder how experts feel about these kinds of arguments. But the evidence is very difficult to refute without further kinds of evidence.

I want to say one word about the distinction between the way the evidence is evaluated by Bayesian statisticians versus frequentist statisticians, which are another variety. Namely, a Bayesian would be willing to say that given the word usage in the *Federalist Papers*, there is a 99.9% chance that it's written by Madison. Whereas, the frequentist would say, "That makes no sense. Either the paper was written by Madison, or it was not written by Madison. We should not ascribe a probability to the concept of something that is either true or not true."

Instead, we could make assertions like the following. Randomly selecting a collection of 1,000 words from a person's writing who generally has 6 "upons" per 1,000 would have a probability of such-and-such occurring by chance alone to have a document that has no use of the words "upon." That kind of reasoning would be agreed to by both camps.

Using data and statistical analysis is obviously a central part of our world today, but I will argue that it will become an even more prominent part of our world in the future. The main reason is the computer, the continuing development of computer technology. With the computer, it's now possible to deal with large databases, and we can use techniques that were previously computationally impossible.

Some techniques, for example, involve simulation as a means to understand a collection of data, where you simulate things, and see what happens. Often, these methods are computationally intensive. Consequently, as we become increasingly able to have more computer power applied to them, then these methods become concomitantly increasingly valuable.

In these lectures, by the way, we've shown several times examples where we've simulated data to demonstrate the efficacy of, for example, an estimator. Remember in the German tank example, we said, "How are we going to estimate how many German tanks there were?" and we simulated, using what we proposed as a method of estimation, in order to get a sense of how effective that estimator really was.

There are methods called the *Monte Carlo method*—referring to the gambling place—of using random processes by a computer to determine very specific things. For example, you can randomly think about simulating throwing a needle on the floor in a method of doing it millions of times in order to determine the value of the constant pi (π). It's a very strange thing, and I won't explain it here, but that's an example of a Monte Carlo kind of method, where randomness and computer simulation can lead to a definite result.

At home, you can now do amazing statistical analyses essentially instantly. You can perform statistical experiments on your [computer] desktop that would make any early 20th century statistician, like Fisher, just drool. They had to do these elaborate computations. Early textbooks on statistics would emphasize methods for reducing computation because that was one of the obstacles to applying the logic of statistics to real situations. Now, of course, those kinds of techniques are not so important.

In this final lecture, I'd like to make some observations about the statistical enterprise altogether. First, there's often a difference between statistical knowledge and understanding based on deeper principles. We saw this example when we discussed the difference between Kepler and Newton in their relationship to the elliptical orbit of the planets. Kepler was doing a statistical model-fitting technique—that is, he had data and found that an ellipse fit the data quite well. Newton, on the other hand, had a theoretical concept about the law of physics from which elliptical orbits followed as a required consequence.

I want to give a more down-to-earth example that occurred when I was serving on a committee, which dealt with preparing future mathematics teachers. One of the members of this committee came into one of the meetings and said, "Well, I went to a 6th grade mathematics class where they were preparing students to take one of these assessment tests." These tests are so prevalent now, these high-stakes assessment tests, and they take them very seriously, the preparation for these tests. The woman on the committee explained what the teacher had done.

She said, "The teacher was telling the students how to do well on these multiple-choice tests, and this is what she explained. She told the students to read the word problem and to look for the numerals, the numbers, and then look at the word that's right after each of the numbers. If the word after each number is the same word, then you add or subtract the numbers, and look in the multiple choice answers to see if any of them occurs. Whereas, if the word following the numbers is a different word, then you multiply or divide the numbers, and then see if any of the multiple choice answers have the answer."

This was what was being taught as a way to succeed. Think about it. I thought this is crazy. Actually, it works because look at the following kinds of word problems. If you have 7 books on a shelf and 8 books on another shelf, then you add them up to see how many books you have altogether. Whereas, if it says, you have 7 books on a shelf and you have 5 shelves, then you multiply them to get the right answer.

It's a heuristic. It's a statistical method by which to come to the conclusion of passing the test, but this is such a clear example of a misuse of statistical methods. Obviously, the goal of the test is to ascertain some quality that is now being perverted by teaching people how to, so to speak, get around the actual goal of this test.

Another common misuse of statistics involves the blind application of statistical tests. I think it's fair to say that statistics is used more than it's understood. Statistics provides a whole collection of methods to analyze data, and methods to educe evidence for various possible inferences from data. But we need to understand the mathematical and the probabilistic logic behind the statistical techniques. If we do, we can be very clear about what the particular statistical test or statistical statement means, and also what it does not mean.

If we attempt to reduce the discussion to some sort of formulaic adherence to following tests, it's extremely likely that we're going to get misleading results or, often, nonsensical arguments. For example, if we just thought about taking averages, if we took the mean of the wealth of all the graduates of Lakeside School, it gave us a very dubious picture, when every person got millions of dollars. That kind of thing is just a blind application of something.

I asked a friend of mine, who is a professor of microbiology, "How is statistics actually used in your laboratory?" She said, "I'll tell you how my graduate students use statistics. They gather data, they enter it into Excel, and then they apply every statistics test in Excel, and if any of them come out with a p value of less than .05, they declare that to be victory, and report their result."

We've seen over and over again statistical reasoning is sufficiently subtle, and sufficiently prone to these counterintuitive examples—which we've seen throughout this course—that actually understanding the underlying logic is absolutely necessary in order to have confidence in the result. We've seen the example of choosing a good hospital. We saw that you had to think more deeply than just the first statistics. Or we saw in the lecture about law, evaluating the evidence against this cabbie who may have been driving a blue or a green cab. Remember that we had to look more deeply in order to see what the implications of the evidence really were.

One of the basic strategies of statistical inference, you recall, is the logic of hypothesis testing. Remember, this is familiar to us, since it's really like our judicial system. We're to assume the person is innocent until the evidence tells us otherwise. That is, in hypothesis testing, we suppose hypothetically that the world is such-and-such a way, and then we determine whether it would be a rare event to find the data that we actually find. Remember, that was the strategy of hypothesis testing.

But hypothesis testing has issues of its own—for example, the arbitrariness of the level of rarity that we deem statistically significant. How rare is rare? This and other issues have to be understood in the context of each application that we try. Anybody can blindly assert that if you get a p value of .049—which is less than this threshold of .05—that means it's a better result, then that implies that the evidence is statistically significant and, therefore, the null hypothesis is false. Whereas, if the outcome gives a p value of .051, that person tells us that the null hypothesis is true. That person just doesn't understand the reasoning behind the hypothesis test, and the probabilistic nature of the appropriate conclusions from statistical inferences.

The Bayesian strategy of updating an *a priori* concept of probability of the world, viewing the world as something that we don't know, and that each part of it has a probability that needs to be updated as we gather more evidence, this—in some sense—gives a more continuous way of interpreting data. Some people find it very persuasive.

The persuasive strength of a statistical argument requires a clear understanding of the statistical reasoning, of the context of the situation, and the details about the study or the data that all allow us to interpret the meaning of the statistical data and arguments with conviction.

Unless we really have a high confidence that a survey, for example, was conducted appropriately—that the questions weren't slanted; that the sample wasn't biased; that the interpretation of the answers actually did capture the issue that we were truly interested in—then the statistical result is it's a piece of evidence, all right, but it may not be as strong as reported.

We saw this example in the *Literary Digest* poll for the Presidency of 1936, and maybe even more startlingly in the Ann Landers example. Those were cases where completely inappropriate conclusions seemed to come from a kind of a survey. Good statistical results and inferences are far superior to anecdotal evidence on an issue. That's true, but we need to be critical consumers of statistical evidence.

So, in a fundamental sense, statistical knowledge is, by its very nature, an admission of ignorance. If we knew all the factors that create happiness, we wouldn't have statistical studies that assert that 50% of people with this characteristic have this result. What it means is that we don't have the whole story yet. We're not able to say that if we do this and we do that and we do the other thing then the economy will definitely get better. Instead, we assert that the probability of its becoming better will increase.

Statistics, it's an admission of ignorance. It's extremely useful in dealing with that basic reality of our ignorance, but expecting surety from statistical results just is a misunderstanding of the fundamental nature of the probabilistic processes with which statistical methods are all involved. Often when we know something statistically, that knowledge actually gives us two pieces of information. One is the indication of reality given by the statistics. The second is an indication of what questions we need to ask to gain a deeper understanding of what's going on.

For example, if we learn that a higher percentage of people who have high cholesterol develop heart trouble than those with lower cholesterol—then we're led to investigate the mechanism by which high cholesterol causes heart problems. In some cases, the underlying reason for the correlation may not be known. In those cases, if we're talking about medicines, it's a question about whether to treat the symptom or not.

In a very fundamental way, much of the human experience—it seems to me—is centered on the quest to derive meaning from the information we receive about the world. In all realms, there are challenges, but when the information comes in the form of data—particularly numerical data—we meet special challenges. So, statistics is a collection of profoundly powerful methods for understanding our world with more detail and more meaning.

We've seen statistics as having two parts. First, there's organizing, describing, and summarizing a collection of data when we know all the data. Second, there's inferring information about the whole population when we have data only about a sample of the population.

So, the guiding themes of describing all the data on the one hand and making statistical inferences on the other give us the framework on which the whole of statistics is built. Statistical methods help us to put structure on complex collections of information. We've seen how to think about the whole distribution of data, rather than being content with one-number summaries, like the mean.

We've developed strategies by which we can reasonably infer consequences from statistical and probabilistic situations. The goal here is to answer the questions, "How close?" and "How confident?" How close is our estimate of an unknown feature of the population? How confident are we that that estimate is, in fact, close?

Statistics is really a powerful tool to help us understand the world. In fact, it's so powerful that I think we should end the course with a statistic about statistics itself. That is, among people who learn something about statistics, 100% appreciate our world with more clarity.

It's been my privilege and my pleasure to deliver these lectures. I feel honored that you've listened. Thank you. Bye for now.

Timeline

1532 .... First weekly data collected on deaths in London.

1539 .... Beginning of official data collection on baptism, marriages, and deaths in France.

1608 .... Beginning of the Parish Registry in Sweden.

1662 .... John Graunt publishes *Natural and Political Observations Mentioned in a Following Index and Made upon the Bills of Mortality*, which initiated the idea that vital statistics can be used to construct life and mortality tables for the relevant population.

1666 .... First modern national demographic census (conducted in Canada).

1689 .... Jacob Bernoulli publishes the law of large numbers, a mathematical statement of the fact that when an experiment is repeated a large number of times, the relative frequency with which an event occurs will equal the probability of the event.

1693 .... Edmund Halley publishes *Estimate of the Degrees of Mortality on Mankind*, which contained the mortality tables for the city of Breslau, Poland. It was one of the earliest works to relate mortality and age in a population and was highly influential in the future production of actuarial tables in life insurance.

1713 .... Nicholas Bernoulli edits and publishes *Ars Conjectandi* (*The Art of Conjecture*), written by his uncle, Jacob

Bernoulli, in which the work of others in the field of probability was reviewed and thoughts on what probability really is were presented.

1728 .... Sir Isaac Newton publishes *The Chronology of Ancient Kingdoms Amended*, in which he gives a 65% confidence interval for the length of a king's reign.

1733 .... Abraham de Moivre publishes an account of the normal approximation for the binomial distribution for a large number of trials. This work improves on Jacob Bernoulli's law of large numbers. The account will be included in the 1756 edition of de Moivre's *The Doctrine of Chances*, a treatise on probability first published in 1718.

1735 .... The beginning of demographic data collection in Norway.

1746 .... Publication in France of tables based on mortality data.

1749 .... Sweden's first complete demographic census.

1753 .... Austria's first complete demographic census.

1766 .... First publication of Sweden's mortality tables.

1769 .... Denmark and Norway's first complete demographic census.

1790 .... Giammaria Veneziano Ortes publishes *Reflessioni sulla popolazione delle nazioni per rapporto all'economia nazionale* (*Reflections on the Population of Nations in Respect to National Economy*).

1790 .... America's first federal demographic census.

1801 .... Britain's first complete demographic census.

1801 .... France's first complete demographic census.

1810 .... Pierre-Simon Marquis de Laplace publishes a fairly general statement of the central limit theorem.

1825 .... Benjamin Gompertz publishes *On the Nature of the Function Expressive of the Law of Human Mortality*, in which he uses logarithmic regression to show that the mortality rate increases exponentially as people age.

1827 .... Pierre-Simon Marquis de Laplace publishes a paper on multiple regression analysis with applications to lunar tides and the atmosphere.

1829 .... Belgium's first complete demographic census.

1834 .... Establishment of the Statistical Society of London (later the Royal Statistical Society).

1835 .... Adolphe Quetelet publishes *Sur l'homme et le développement de ses facultés, essai d'une physique sociale*, in which he presents his conception of the average man as the central value about which measurements of a human trait are grouped according to the normal distribution.

1837 .... Simeon Denis Poisson publishes *Recherches sur la probabilité des jugements en matière criminelle et matière civile*, which introduces the expression *law of large numbers* and in which the Poisson distribution first appears.

1837 .... Public data collection of the demographic statistics in England. Establishment of the Registrar General Office.

1839 .... Organization of the American Statistical Association.

1846 .... Verhulst publishes a nonlinear differential equation describing the growth of a biological population, which he deduced from data. The equation predicts that population growth is limited by forces that increase with the square of the rate at which the population grows, rather than being unlimited exponential growth.

1853 .... Augustin-Louis Cauchy presents an outline of the first rigorous proof of the central limit theorem.

1853 .... Adolphe Quetelet organizes the first international statistics conference.

1861 .... Italy's first complete demographic census.

1867 .... Pafnutii Lvovich Chebyshev publishes a paper, *On Mean Values*, which uses Bienaymé's inequality to give a generalized law of large numbers.

1869 .... Establishment of the *Société de Statistique de Paris* (the Statistical Society of Paris).

1885 .... Establishment of the International Statistical Institute in the Netherlands.

1887 .... Pafnutii Lvovich Chebyshev publishes *On Two Theorems*, which gives the basis for applying the theory of probability to statistical data, generalizing the central limit theorem of de Moivre and Laplace.

1889 .... Francis Galton publishes *Natural Inheritance*, in which he presents a summary of the work he had done on correlation and regression that included the idea of regression to the mean, discovered through his 1875 experiments with sweet peas.

1892 .... I.V. Sleshinsky publishes the first complete rigorous proof of the central limit theorem, based on the outline by Cauchy.

1893 .... Karl Pearson begins the publication of 18 papers entitled *Mathematical Contributions to the Theory of Evolution*, which contain his most valuable work in the form of contributions to regression analysis, the correlation coefficient, and the chi-square test of statistical significance. This work lasts through 1912.

1897 .... George Udny Yule publishes *On the Theory of Correlation*, in which he begins the development of his approach to correlation via regression with a conceptually new use of least squares that would later dominate applications in the social sciences.

1901 .... Publication of the first issue of *Biometrika*, a journal founded by Karl Pearson and Francis Galton.

1908 .... William Sealy Gosset, under the pseudonym "Student," publishes the t-distribution as the sampling distribution of the mean when the population variance is unknown.

1921 .... Ronald A. Fisher introduces the concept of maximum likelihood. In 1922, he would redefine statistics such that its purpose was the reduction of data. Fisher identified three fundamental problems: (1) specification of the kind of population from which the data came, (2) estimation, and (3) distribution.

1930 .... Establishment of the Institute of Mathematical Statistics and the appearance of *Annals of Mathematical Statistics* in the United States.

1931 .... Establishment of the Indian Statistical Institute.

1933 .... Jerzy Neyman and Egon Pearson publish *On the Problem of the Most Efficient Tests of Statistical Hypotheses* and *The Testing of Statistical Hypotheses in Relation to Probabilities a Priori*, capping five productive years of research on statistical hypothesis testing.

1935 .... Ronald A. Fisher publishes the first edition of *The Design of Experiments*, which revolutionizes the use of statistics in agriculture.

1937 .... George W. Snedecor and William G. Cochran publish *Statistical Methods.*

1946 .... Harald Cramer publishes *Mathematical Methods of Statistics*, which joins the science of statistical inference with the theory of classical probability and was reprinted as recently as 1999.

1947 .... Ronald A. Fisher publishes *Statistical Tables.*

1966 .... Foundation of the Working Party on Statistical Computing, which published guidelines for program development, descriptions, and code for statistical programs.

1967 .... W. J. Hammerle publishes *Statistical Computations on a Digital Computer*, the first textbook devoted to statistical computing.

1972 .... The American Statistical Association forms a section for statistical computing. In the following years, the use of computers in statistics will allow statisticians to generate, collect, organize, and analyze larger data sets and increase the complexity of the models fitted to the data. Displays become more impressive, using color and perspective. In the realm of hypothesis testing, permutation testing undergoes a resurgence. Computing power allows statisticians to develop theory using Monte Carlo simulation studies.

Glossary

analysis of variance (ANOVA): A procedure of statistical analysis by which differences in means of two or more groups can be assessed after eliminating variance that is due to other factors.

Bayesian statistics: The view in which probability is interpreted as a measure of degree of belief. In this view, the concept of probability distribution is applied to a feature of a population, such as the population mean, to indicate one's belief about possible values of that feature. The principal result of experiments is to update such a probability distribution, indicating a change in belief. The Bayesian viewpoint is in contrast to the frequentist view.

bias: The extent to which the statistical method used in a study does not estimate the quantity to be estimated or may not test the hypothesis to be tested.

binomial distribution: The probability distribution of the number of successes in n Bernoulli trials. For a series of events to be considered Bernoulli trials, they must satisfy three conditions: (1) the trials are independent of each other, (2) each trial has exactly two possible outcomes, and (3) the probability associated with each outcome is constant throughout all of the trials.

box plot: A graphical display for numerical data that shows the maximum and minimum values, the median, and the quartiles of the data.

central limit theorem: Statistical theorem that states the following: Starting with almost any distribution (such as a Poisson distribution, a binomial distribution, or a uniform distribution) with a finite standard deviation σ, if we take many samples of size n, the distribution of the average values of the samples will be approximately a Gaussian distribution (assuming n is large) with the same mean as the original distribution and with standard deviation $\frac{\sigma}{\sqrt{n}}$.

chi-square distributions: A family of distributions that take only positive values and are skewed to the right. Each chi-square distribution is specified by its degrees of freedom. The higher the degrees of freedom, the more skewed the distribution is. The chi-square family of distributions occurs often in hypothesis testing about categorical variables.

chi-square test for independence: A process used to test the hypothesis that two categorical variables have no relationship. The test statistic that is calculated has a chi-square distribution.

confidence interval: A range of values, constructed from information obtained from a sample of the population, that is believed, with a specified probability, to contain the value of the population parameter.

correlation coefficient: The quantification of the strength of linear association that exists between two numeric variables. The correlation coefficient takes values between –1 and 1, where negative correlations mean that as the value of one variable rises, the other falls, and positive correlations mean that the values of the two variables rise together. Values of the correlation coefficient near 1 or –1 indicate a strong linear relationship between the two variables. Values near 0 indicate no linear relationship between the two variables.

dispersion: The variation among values when the data values in a sample are not all the same.

estimator: A statistic, calculated based on the information from a sample, that is used to estimate the value of a parameter associated with the population from which the sample was selected.

event: An outcome or set of outcomes from a random process.

expected value: The average outcome that might be expected from a long run of trials of a probabilistic event.

experimental design: Procedures and planning used in an experimental study. In general, these procedures are designed to reduce bias, promote replication, use randomization in order to initiate study of causality, and ensure appropriate sample size.

extrapolation: The process of using the data to make estimates about values that lie beyond the range of the existing data.

factor analysis: A set of statistical procedures used to analyze multivariable data when many variables are known about the subjects. The underlying principle behind factor analysis is that variables that are highly correlated with each other are grouped together and separated from variables that are not highly correlated with the group. Each group represents a factor, thought to be a single underlying construct.

five-number summary: A numerical summary of data that includes the minimum and maximum values, the median, and the upper and lower quartiles. The five numbers divide the data into four groups, each containing the same number of data points. Often used to describe data that have skew.

frequentist statistics: The view in which probability is defined in terms of long-run frequency or proportion in outcomes of repeated experiments. The concept of probability is applied to outcomes of actual or hypothetical experiments because there is a random element to those. But in the frequentist view, probability is not used as a measure of knowledge or belief of the possible values of a quantity, such as the true population mean, that does not have a random element. The frequentist viewpoint is in contrast to the Bayesian view.

Gaussian distribution: See **normal (Gaussian) distribution**.

histogram: A graphical display for numerical data in which vertical bars show the number of observations that have a value between the values given on the x-axis at the base of the bar.

hypothesis test: The process of assessing whether observed data are consistent with some claim about the population in order to determine whether the claim might be false.

independent events: Two events are independent if knowing the outcome of one tells us nothing about the other. There is no relationship between the two events.

interquartile range (IQR): A measure of spread. The difference between the upper quartile and lower quartile. Also used in rules of thumb for identifying outliers.

lurking variable: A variable that has an important effect on the relationship among variables considered in a study but that is not, itself, considered in the study.

mean: A measure of the location of the center of numerical data. Also called the *arithmetic average*. It is computed by summing the values of the data and dividing by the number of data points. Conceptually, it is the balance point of the data when they are represented by a line plot. Because the mean is not particularly resistant to outliers, it is used mainly when the data have a roughly symmetric distribution.

median: A measure of the location of the center of numerical data. Once the data are ordered by their value, the median is the value taken by the data point that is in the middle, such that there are the same number of data points larger than the median as smaller than the median. If there is an even number of data points, then the median is the average of the values of the two in the middle. The median is also the 50th percentile and the second quartile. Because the median is particularly robust to outliers, it is used when the data are skewed or contain outliers.

Monte Carlo method: A numerical modeling procedure that makes use of random numbers to simulate processes that involve an element of chance. In Monte Carlo simulation, a particular experiment is repeated many times with different randomly determined data to allow statistical conclusions to be drawn.

nonparametric test: Ill-defined term used generally to describe processes for inference that may be used either when the assumptions underlying parametric procedures, such as chi-square and one- and two-sample tests, are not met or when responses are difficult to quantify or contain rankings rather than meaningful numerical values.

normal (Gaussian) distribution: A family of single-peaked, symmetric probability distributions described as bell shaped. It is the distribution associated with errors in measurement, with heights and weights, and with standardized test scores, for example.

null hypothesis: A proposition or set of propositions to be tested.

observation: The value associated with one member of a sample.

one-sample test for means: A process for testing the hypothesis that the mean value of some quantitative aspect of a population has a particular value. The test statistic exhibits a roughly t-distribution when the standard deviation of the value in the population is not known.

one-sample test for proportions: A process for testing a hypothesis about the percent of members of a population who have a particular characteristic or opinion. The test statistic has a roughly normal distribution.

outlier: A data point with value that differs markedly from the rest of the values in the data set.

p-value: In a hypothesis test, the probability of obtaining the results that were obtained from a sample or results more unusual if the null hypothesis represents the truth about the population.

parameter: A numerical value about data that is calculated from the values of a population.

percentiles: The percentiles are the observations that divide the data into 100 groups, each with the same number of observations. For example, scoring in the 85th percentile on the SAT means that one has outscored 85% of those tested.

Poisson distribution: A right-skewed probability distribution that describes the number of occurrences of an event in a given time period.

population: A population is any entire collection of people, animals, plants, or things from which we may collect data. It is the entire group in which we are interested and that we wish to describe or draw conclusions about.

power: The power of a hypothesis test is the ability of the test to accurately reject the null hypothesis when the null hypothesis is, indeed, false. One wants tests to have high power. However, as the power of a test increases, so does the probability of a type I error, that is, the rejection of the null hypothesis when it is actually true. The statistician must find a reasonable balance between power and the probability of a type I error.

probability distribution: A probability distribution is a table, function, or graph that assigns a probability to each possible outcome.

quartiles: The quartiles are the values of the data that divide the observations into four equal-sized groups. To find the quartiles, list the values of the data in order from smallest to largest. The second quartile (median) is the observation in the middle. The first quartile is the observation that divides the lower half of the data, between the minimum and the median, into two equal-sized groups. The third quartile is the observation that divides the top half of the data, between the maximum and the median, into two equal-sized groups.

regression analysis: A statistical process by which a model is created that predicts the value of a response variable through an equation using the values of one or more explanatory variables.

residual: The difference between the actual (observed) value of a response variable and that calculated from a regression equation.

sample: A subset of a population that is used to infer information about the population.

sample mean: The value of the mean of a sample.

sampling bias: Error that is introduced in a statistical study by the method of sampling. For example, the use of voluntary sampling, such as online polls, introduces bias because the respondents tend to be those who are passionate about the topic, rather than a random sample of people with all types of opinions.

sampling distribution: The theoretical distribution of the statistic calculated from a sample. The generation of this distribution is based on the calculation of the statistic from every possible sample from the population.

scatter plot: A two- or three-dimensional graph in which each axis represents one variable that is associated with an observation. Used in regression analysis as a visual display of patterns that may exist among variables in the data.

significance level: In a hypothesis test, a prespecified value at which the null hypothesis may be rejected. Sometimes used to describe the p-value of a hypothesis test, that is, the probability of obtaining the value that was obtained from a sample if the null hypothesis about the population from which the sample was selected is true.

simple random sample (SRS): A sample of a population that is chosen in such a way that each member of a population has an equal chance of being selected.

skewness: The lack of symmetry exhibited by a distribution. The direction of skew, left or right, tells the direction of the tail that causes the lack of symmetry.

standard deviation: The most commonly used measure of dispersion (spread) for numerical data. It is the square root of the variance. Like the variance and the mean, its calculation is not resistant to outliers and extreme skew.

standard error: The standard deviation of the sampling distribution of a statistic.

statistic: A numerical value about data that is calculated from the values of a sample.

stochastic: A synonym for random; the adjective applied to any phenomenon obeying the laws of probability.

stratified sample: A method of sampling by which the population is first divided into groups, or strata, based on common characteristics, such as gender or income. If a random sample is then selected from each group, the term *stratified random sample* may be used.

t-distribution: The t-distribution is the theoretical distribution of a sample mean calculated from a sample taken from a population whose standard deviation is not known. Its shape is roughly symmetric and similar to that of a normal distributed variable, but the tails are thicker.

two-sample test for means: A process for testing the hypothesis that two different populations have the same mean. The calculated test statistic is theorized to have a t-distribution.

two-sample test for proportions: The process for testing the hypothesis that two different populations share the same value for a binomial process (such as a yes-no question). The calculated test statistic has a roughly normal distribution.

type I error: Rejection of the null hypothesis when it is true.

type II error: Acceptance of the null hypothesis when it is false.

uniform distribution: A distribution in which every possible value is equally likely. The histogram of a uniform distribution has all of the bars the same height.

variance: A measure of dispersion (spread) for numerical data. It is roughly the average squared distance of the data values from the mean. It is calculated by summing the square of the differences between the data and the mean. To calculate a population variance, one divides by the number of elements in the population. To calculate the sample variance, one divides by one fewer than the number of observations in the sample.

Biographical Notes

Bayes, Thomas (1701–1761). British nonconformist minister. Little is known about Bayes's life save that he was the son of a nonconformist minister, educated at Edinburgh University, and a member of the Royal Society. His major contribution to the field of statistics was the work he did on the inverse probability problem. At the time, the calculation of the probability of a number of successes out of a given number of trials of a binomial event was well known. Bayes worked on the problem of estimating the probability of the individual outcome from a sample of outcomes and discovered the theorem for such a calculation that now bears his name.

Bernoulli, Jacques (often called Jacob or James) (1654–1705). Professor of mathematics at Basel and a student of Leibniz. He formulated the law of large numbers in probability theory and wrote an influential treatise on the subject.

Cauchy, Augustin-Louis (1789–1857). French mathematician and engineer. Professor in the Ecole Polytechnique and professor of mathematical physics at Turin. Cauchy worked in number theory, algebra, astronomy, mechanics, optics, and analysis. His contribution to statistics was the production of the outline of the first rigorous proof of the central limit theorem in 1853, in the course of a controversial debate during meetings of the Academy of Sciences and in the pages of its journal with Irenée-Jules Bienaymé (1796–1878). The debate started as a result of a critique made by Cauchy of the work of Laplace. Bienaymé, a student of Laplace, took exception to the criticism of his mentor, on whose work much of Bienaymé's was based, and a debate ensued. Although Cauchy only sketched his proof, I. V. Sleshinsky was able to fill in the details and missing steps. He produced a complete, rigorous proof of the central limit theorem based on the outline by Cauchy in 1892.

Chebyshev, Pafnutii Lvovich (1821–1894). Russian mathematician, founder of the St. Petersburg School of Mathematics. The culmination of his career of study in probability and statistics occurred in 1887, with his use of the method of moments to prove the first version of the central limit theorem for sums of independent but not identically distributed variables.

Cox, Gertrude Mary (1900–1978). American statistician and administrator. Cox's main contributions to the field of statistics were in the areas of experimental design and analysis of psychological data. In addition, in 1949 she became the first woman elected to the International Statistical Institute. Cox was the first head of the department of experimental statistics at North Carolina State University. She was a founding member of the Biometrics Society and editor of the journal *Biometrics* for 10 years.

Cramer, Harald (1893–1985). Swedish mathematician and statistician. Chair of the actuarial mathematics and mathematical statistics department and, later, president of Stockholm University, Cramer served as chancellor of the Swedish university system. He wrote several seminal books that expressed probability theory in a manner more useful in its application to statistical theory than had previously been articulated. Working as an actuary with the Svenka Life Insurance Company early in his career led Cramer to investigate stochastic processes as they related to insurance. His text, *Collective Risk Theory*, is concerned with the progress over time of monetary funds, with inputs, such as premiums and interest, and outputs, such as claims, as special cases of general stochastic processes.

Deming, W. Edwards (1900–1993). American statistician and quality-control expert. Trained as a physicist, Deming became interested in statistics while working at the U.S. Department of Agriculture. He then took a post at the U.S. Bureau of Census as the Head Mathematician and Advisor in Sampling. He is credited with importing the replicate subsampling method from India, which forms part of the national sampling plan used by the U.S. Bureau of Census and by polling corporations, such as Gallup. After World War II, Deming was assigned to General MacArthur's Supreme Command of the Allied Powers in Tokyo. While there, Deming undertook a systematic education of quality-control principles and techniques in the Japanese workforce. The Japanese attention to quality control as introduced to them by Deming is credited as the primary force behind that country's emergence as an industrial leader among nations.

de Moivre, Abraham (1667–1754). French-English mathematician. Born in France and educated at the Sorbonne in mathematics and physics, de Moivre, a Protestant, emigrated to London in 1688 to avoid further religious persecution. A future fellow of the Royal

Society of London, de Moivre supported himself in England as a traveling mathematics teacher and by selling advice in coffee houses to gamblers, underwriters, and annuity brokers. De Moivre is recognized in statistics as the first to publish an account of the normal approximation to the binomial distribution. In fact, some of de Moivre's methods are so ingenious as to be shorter than modern demonstrations of solutions to the same problems.

Fisher, Ronald Aylmer (1890–1962). British statistician. Trained in mathematics and physics, Fisher is known as the father of modern statistical methods. Through correspondence with W. S. Gosset, Fisher was the first to derive the general sampling distribution of the correlation coefficient. His major contributions to statistics were in the area of design of experiments. He introduced the concept of randomization and the process of analysis of variance (ANOVA) now widely used by statisticians. He was a fellow of the Royal Statistical Society and was elected to the American Academy of Arts and Sciences, the American Philosophical Society, the International Society of Haematology, the National Academy of Sciences of the United States, and the Deutsche Akademie der Naturforscher Leopoldina. He was awarded honorary degrees from many institutions, including Harvard University (1936), University of Calcutta (1938), University of London (1946), University of Glasgow (1947), University of Adelaide (1959), University of Leeds (1961), and the Indian Statistical Institute (1962). Fisher was knighted in 1952.

Galton, Francis (1822–1911). British explorer and anthropologist. Cousin to Charles Darwin. He was the first to calculate a quantitative value for correlation and a pioneer of the use of the variable r for correlation coefficient, although his calculation differs from that used by modern statisticians. He was the first person to document the phenomenon known as *regression to the mean*, which he discovered through experiments with sweet peas. His ideas strongly influenced the development of statistics, particularly his proof that a normal mixture of normal distributions is itself normal. Galton may be described as the founder of the study of eugenics. His principal contributions to science consisted of his anthropological inquiries, especially into the laws of heredity. In 1869, in *Hereditary Genius*, he endeavored to prove that genius is mainly a matter of ancestry via the application of statistical methods.

Gauss, Karl Friedrich (1777–1855). German mathematician and astronomer, nicknamed the "Prince of Mathematicians." His mathematical work included the concept of a distribution of errors that originally was known as the *error distribution* and later became known as the *Gaussian distribution*, or the *normal distribution*.

Gosset, William Sealy (1876–1937). British chemist who, while working at the Guinness Brewery in Dublin, Ireland, began a study of statistical methods as applied to small samples. Asked by the brewery to investigate the relationship between the quality of materials, barley and hops, for example, and production conditions on the product, beer, the corporation required him to publish his results under a pseudonym to preserve the anonymity of the brewery. Gosset chose the name "Student" under which to publish his results about the derivation and use of a t-distribution in inference, leading to its being referred to as the *Student's t distribution*. In later work at the brewery, Gosset would come to support the use of a balanced design in agricultural applications, rather than either of the two available competing designs. Unfortunately, Gosset would pass away before this disagreement could be resolved.

Laplace, Pierre-Simon Marquis de (1749–1827). French mathematician and astronomer. Professor at Ecole Normale and Ecole Polytechnique. Primarily known for his contributions to calculus, analysis, and physics, toward the end of his life, he turned to research in statistics and obtained a fairly general statement of the central limit theorem in 1810. Between 1818 and his death, Laplace investigated multiple regression analysis as related to lunar tides and the atmosphere and published a comparison of absolute versus least-squares deviations and their use in regression analysis and the notion of a sufficient statistic.

Markov, Andre Andreevich (1856–1922). Russian mathematician. Member of the St. Petersburg Academy of Science. Markov was a student of Chebyshev and spent most of his career studying probability distributions, random variables, the weak law of large numbers, and the central limit theorem. Markov's significant contribution to probability theory was the introduction of the concept of a Markov chain as a model for studying the behavior of random variables. One example of a Markov chain is known as a simple random walk. In a random walk, each direction in which a "man" may step is assigned a probability. The paths that may occur and their assigned probabilities make up the "behavior" of this particular random variable. In modern statistical practices, Markov chains are used in conjunction with Monte Carlo methods to solve problems that are analytically complicated by generating suitable random numbers and observing the fraction of those numbers that obey some property or properties.

Mendel, Johann Gregor (1822–1884). Czech monk. Mendel was the first to apply statistical methods to biology in his calculation of ratios of genotypic structures. The application of his work to the field of genetics was not recognized until the 1930s, 50 years after his death. The validity of Mendel's most famous work, on the hybridization of peas, has come under question. Fisher initially concluded that Mendel's description of experimental design was correct but that the data lacked an appropriate amount of random variation and were likely fabricated or sanitized. Later authors have exonerated Mendel based on his use of sequential procedures and the inclusion of meteorological data.

Moore, David. Prolific and lucid author of statistics textbooks. His work was influential in redefining the common presentation of statistics in colleges, de-emphasizing the mathematical theory, and focusing on real data.

Newton, Sir Isaac (1642–1727). English mathematician and scientist known for the discovery of the law of gravity and as one of the fathers of calculus. Within the field of probability, Newton is known for his proof of the binomial theorem. There is also evidence that he gave thought to the variability of the sample mean, the basis for the central limit theorem. In his last work, *The Chronology of Ancient Kingdoms Amended*, published posthumously in 1728, Newton estimated the mean length of a king's reign to be between 18 and 20 years. In fact, the mean reign was 19.1 years, and the standard deviation of his sample was 1.01 years; thus, Newton's range of 18 to 20 years roughly corresponds to a 65% confidence interval (Johnson and Kotz).

Neyman, Jerzy (1894–1981). Polish statistician. Reader at University College in London and founder of the department of statistical sciences at the University of California at Berkeley. Elected to the International Statistical Institute and the U.S. National Academy of Sciences. Neyman's work in statistical inference leads some to call him the father of modern statistical methods. In a paper co-authored with Karl Pearson, Neyman explained the logical foundation and mathematical basis for the theory of hypothesis testing. Through his work, the theory of confidence intervals was developed from the theory of hypothesis testing. Neyman also contributed to innovative and precise use of statistics in fields ranging from agriculture and astronomy, biology, and social insurance to weather modification.

Nightingale, Florence (1820–1910). British nurse. Commonly known as the "Lady of the Lamp" for her work as a nurse for British troops during the Crimean War, Nightingale saved more soldiers through the use of statistics than medicine. Utilizing both statistical methods and innovative graphical techniques, she was able to convince the British army of the importance of hygiene and sanitation in hospitals, which led to widespread army hospital reform.

Pareto, Vilfredo Federigo Samaso (1848–1923). French-Italian engineer and economist. His contributions to statistics include work on interpolation and fitting curves to data and actuarial calculations in insurance and pensions. However, Pareto's most significant contribution to statistics was in his discovery of the first stable probability distribution other than the Gaussian (normal) distribution, named the *Pareto distribution* in his honor. Not only does the Pareto fit naturally arising situations, such as income distribution, but it also has theoretical applications. When conditions for the central limit theorem do not apply because the population distribution has heavy tails and, therefore, does not have finite variance, a modification of the central limit theorem may be applied if the behavior of the tail of the population distribution has roughly a Pareto distribution.

Pearson, Karl (1857–1936). British mathematician. Chair of the applied mathematics department at London's University College. Influenced by Galton (see above) and Walter Frank Raphael Weldon, a Darwinian zoologist who worked to make biology a more rigorous and quantitative science, Pearson became interested in developing mathematical methods for studying heredity and evolution. Together, the three founded the journal *Biometrika*. Pearson worked out the mathematical properties of both the product-moment correlation coefficient and simple regression used to measure the relationship between two continuous variables. Later in his career, he explored relationships between two categorical variables and mixtures of categorical and continuous variables and developed the chi-square test.

Poisson, Simeon Denis (1781–1840). French mathematician. He published *Recherches sur la probabilité des jugements en matière criminelle et matière civile* in 1837, marking the first appearance of the Poisson distribution, originally found by de Moivre, which describes the probability that a random event will occur in a time or space interval under the conditions that the probability of the event's occurring is very small. Poisson also introduced the expression *law of large numbers*, by which he meant that for a larger number of trials, the proportion of successful outcomes exhibits statistical regularity even if the probability of success does not remain constant. Although we now rate this work as of great importance, it found little favor at the time, the exception being in Russia, where Chebyshev developed Poisson's ideas.

Quetelet, Lambert Adolphe Jacques (1796–1874). Belgian mathematician, astronomer, and meteorologist. In the first recorded attempt to summarize characteristics of the population, Quetelet coined the term *social physics* and used data collected from the national population to describe the average man, *l'homme moyen.* This led Quetelet to the notion that nature was attempting to create the average man as a prototype and that deviations from the average were errors. Working for the government, Quetelet collected and analyzed statistics on crime and mortality and devised improvements in census taking. Influenced by Laplace and Fourier, he was the first to use the normal curve other than as an error law. The distributions of measurements, such as chest circumferences of Scottish soldiers and heights of French conscripts, illuminated the appearance of normally distributed measures in nature and inspired work in fields as diverse as astronomy and physics. At an observatory in Brussels that he established in 1833 at the request of the Belgium government, Quetelet worked on statistical, geophysical, and meteorological data; studied meteor showers; and established methods for the comparison and evaluation of data. His studies of the numerical consistency of crimes stimulated wide discussion of free will versus social determinism. His work produced great controversy among social scientists of the 19th century. Finally, the internationally used measure of obesity, the Body Mass Index (BMI), is derived from the Quetelet index.

Spearman, Charles Edward (1863–1945). British psychologist under whose leadership at University College emerged the "London School" of psychology, distinguished by its rigorous statistical and psychometric approach. Spearman formulated a two-factor theory of human intelligence, in which one factor is common to all mental activities and the other is task specific. He came to identify the first factor through the intercorrelations that existed between scores of subjects on various tests of intelligence. This quantifiable factor has come to be called *g* by cognitive psychologists. Spearman's model was based on a mathematical formulation that laid the groundwork for the statistical methods of factor analysis and contributed to research in test reliability.

Wilcoxon, Frank (1892–1965). American chemist who worked most of his career in industry researching fungicides and insecticides for the Boyce Thompson Institute, the Atlas Powder Company, and the American Cyanamid Company. He was a fellow of the American Statistical Association and the American Association for the Advancement of Science. Wilcoxon studied R. A. Fisher's *Statistical Methods for Research Workers*, which interested him in the application of statistics in experimentation, but through his research, he would seek statistical methods that were numerically simple and more easily understood and applied. Wilcoxon's main contribution to statistics was the development of nonparametric statistical processes, particularly the sign rank tests for two-sample and matched-pairs experiments and his method for multiple comparisons.

Yule, George Udny (1871–1951). Scottish statistician. Secretary, president, and fellow of the Royal Statistical Society. A student of Karl Pearson, Yule made fundamental contributions to the theory of regression and correlation, association between categorical variables, epidemiology, and times-series analysis.

Bibliography

Albert, Jim, and Jay Bennett. *Curve Ball: Baseball, Statistics, and the Role of Chance in the Game*. New York: Copernicus Books, 2003. Introduces the fundamental concepts of statistics through applications to historical baseball. Tackles in detail such issues as the best hitter and hitting streaks.

Barnett, Vic. *Comparative Statistical Inference*. London: John Wiley & Sons, 1973. This book discusses several different approaches to statistical inference and decision making. It compares Bayesian statistics and frequentist statistics and a decision-oriented approach.

Berry, Donald A. *Statistics: A Bayesian Perspective*. Belmont, CA: Duxbury Press at Wadsworth Publishing Company, 1996. An excellent elementary introduction to statistics, with many interesting real examples, most from medicine. The Bayesian approach is used.

Berry, Donald A., and Bernard W. Lindgren. *Statistics: Theory and Methods*, 2nd ed. Belmont, CA: Duxbury Press at Wadsworth Publishing Company, 1996. Introduces theory and application of modern statistics; designed for a year-long course in calculus-based statistics.

Bowerman, B., R. O'Connell, and A. Koehler. *Forecasting, Time Series, and Regression: An Applied Approach*, 4th ed., part II. Belmont, CA: Thomson, Brooks-Cole, 2005. Section of an undergraduate text that offers a good, concise explanation of regression and multiple-regression processes.

Cook, R. Dennis, and Sanford Weisberg. *Applied Regression Including Computing and Graphics*. New York: John Wiley & Sons, 1999. Text for a one-semester undergraduate/graduate course in regression techniques and graphics.

Gonick, Larry, and Woollcott Smith. *A Cartoon Guide to Statistics*. New York: Harper, 1993. A light but very informative view of serious statistics. It often goes right to the heart of a fundamental question.

Gould, Stephen J. *Full House: The Spread of Excellence from Plato to Darwin.* New York: Three Rivers Press, 1996. One of the most popular and prolific science writers of our time, Gould writes equally well about statistics of cancer and of .400 hitters. See part 2, chapter 4, and part 3, chapters 9 and 10.

Heyde, C. C., and E. Seneta, eds. *Statisticians of the Centuries*. New York: Springer-Verlag, 2001. This book contains short biographies of statisticians from the 16th to the 20th centuries.

Huff, Darrell. *How to Lie with Statistics*. New York: W.W. Norton, 1954. This charming little book has been in continuous publication since 1954. It is eminently readable and cheerfully describes methods to mislead with statistics.

Jaynes, E. T., and G. Larry Bretthorst, eds. *Probability Theory: The Logic of Science*. Cambridge, UK: Cambridge University Press, 2003. This book makes the case for the Bayesian approach to statistics, pointing out the difficulties in the orthodox approach. The book is technical.

Johnson, Norman L., and Samuel Kotz, eds. *Leading Personalities in Statistical Sciences: From the Seventeenth Century to the Present.* New York: Wiley, 1997. This book presents biographies of more than 100 statisticians and probabilists of the last four centuries.

Lewis, Michael. *Moneyball: The Art of Winning an Unfair Game*. New York: W.W. Norton, 2003. A delightful story of statistical power in baseball management.

Moore, David S. *Statistics: Concepts and Controversies*, 5th ed. New York: W.H. Freeman and Company, 2001. Informal introductory text on statistical ideas and reasoning; relates these to public policy, science, medicine, sociology, and daily life.

Moore, David S., and George P. McCabe. *Introduction to the Practice of Statistics*, 5th ed. New York: W.H. Freeman and Company, 2005. Collegiate-level introductory text focusing on data analysis, statistical reasoning, and the use of statistics in everyday life.

Paulos, John A. *A Mathematician Plays the Stock Market*. New York: Basic Books, 2003. Engaging, witty stories about what mathematical thinking can disclose about the stock market.

Peters, William S. *Counting for Something: Statistical Principles and Personalities*. New York: Springer-Verlag, 1987. This book illustrates both statistical topics and historical information using short chapters devoted to applications of statistical theory and the personalities who discovered them.

Porter, Theodore M. *The Rise of Statistical Thinking, 1820–1900*. Princeton, NJ: Princeton University Press, 1986. A fascinating historical description of the 19th-century background that led to the innovative development of modern statistics during the early 1900s.

Saari, Donald G. *Chaotic Elections! A Mathematician Looks at Voting*. Providence, RI: American Mathematical Society, 2001. An excellent introduction to the mathematics of voting, written for readers with high school mathematics. Includes an analysis of the American presidential voting scheme.

Salsburg, David. *The Lady Tasting Tea: How Statistics Revolutionized Science in the Twentieth Century*. New York: Henry Holt & Co., 2001. This book presents the history of statistics during the 20th century with humor and insight. Historical anecdotes bring the topics of statistics to life. Readable and enjoyable.

Schmitt, Samuel A. *Measuring Uncertainty: An Elementary Introduction to Bayesian Statistics.* Reading, MA: Addison-Wesley Publishing, 1969. An introduction to Bayesian statistics.

Tufte, Edward R. *The Visual Display of Quantitative Information*. Cheshire, CT: Graphics Press, 1983. Classic text of principles and theories of data graphics, including many historical examples of excellence in visual display.

Wainer, Howard. *A Trout in the Milk and Other Visual Adventures*. Princeton, NJ: Princeton University Press, 2005. Data graphics from a historical perspective in three parts. Part I is the story of the invention and early history of graphing data. Part II illustrates the power of the invention of data graphics, and Part III looks toward the future of visual representations of data.

———. *Visual Revelations: Graphical Tales of Fate and Deception from Napoleon Bonaparte to Ross Perot*. New York: Copernicus, Springer-Verlag, 1997. Descriptions of graphical failures and successes through history. That is, a discussion of the use of misleading graphs and poor execution of visual displays, contrasted with exemplary graphs that led to the discovery of a previously unknown relationship.

Watkins, Ann E., Richard L. Scheaffer, and George W. Cobb. *Statistics in Action: Understanding a World of Data*. Emeryville, CA: Key Curriculum Press, 2004. A textbook that emphasizes student activities.

Software:

Fathom: Dynamic Data Software, version 2. Emeryville, CA: Key Curriculum Press, Key College Publishing, 2005. Fathom is a user-friendly and powerful software system that allows the user to interactively bring ideas in statistics to life. Just by clicking and dragging, it is possible to manipulate data to explore many statistical ideas and illustrate them graphically.

Internet Resources:

Arc Software. University of Minnesota School of Statistics. http://www.stat.umn.edu/arc/software.html. Graphics can be produced in Arc, a free, downloadable statistical analysis tool for regression problems, as described in the book *Applied Regression Including Computing and Graphics* by Cook and Weisberg.

Barton, Paul E. "One-Third of a Nation." Educational Testing Service.
http://www.ets.org/Media/Education_Topics/pdf/onethird.pdf. This 46-page report discusses the fact that one-third of high school students in the United States do not graduate.

The Baseball Archive. www.baseball1.com. This website contains all sorts of data about baseball, including playing statistics, awards, records and feats, history, and salary and payrolls.

Bureau of the Public Debt, U.S. Department of the Treasury. http://www.publicdebt.treas.gov/. This website contains detailed information about the national debt.

"Chance." http://www.dartmouth.edu/~chance/. This website is devoted to helping the teaching of probability or statistics courses. It contains lectures and other materials and a list of links to other sites.

Chance Magazine.
http://www.amstat.org/publications/chance/index.html. This website has accessible articles about statistics and probability and their applications.

Consortium for the Advancement of Undergraduate Statistics Education (CAUSE) (homepage). http://www.CAUSEweb.org. This site has links to many resources about teaching and learning statistics.

Consumer Price Indexes (CPI). Bureau of Labor Statistics of the U.S. Department of Labor. http://www.bls.gov/cpi/home.htm. This site contains a wealth of current and historical CPI data and a very complete description of the CPI and how it is computed.

Fisher, R. A. "Has Mendel's work been rediscovered?" *Annals of Science*, 1 (1936):115–137. http://www.library.adelaide.edu.au/digitised/fisher/144.pdf. This 1936 article by one of the giants of statistics discusses the data in Mendel's famous experiments concerning inheritance of plant characteristics. Fisher's article argues that Mendel's data are too good to be unbiased reporting of real plant growth.

Gehl, B. K., and D. Watson. *Defining the Structure of Jealousy through Factor Analysis.* Los Angeles: Society for Personality and Social Psychology, February 2003. Poster presented at the Society for Personality and Social Psychology Annual Meeting. Available at http://www.psychology.uiowa.edu/students/ gehl/definingjealousy.doc.

Index of Biographies. School of Mathematics and Statistics, University of St. Andrews, Scotland. http://www-groups.dcs.st-andrews.ac.uk/~history/ BiogIndex.html. This website gives biographical information about thousands of noted mathematicians. Both chronological and alphabetical indexes are presented, as well as such categories as female mathematicians, famous curves, history topics, and so forth.

R Development Core Team (2004). R: A language and environment for statistical computing. R Foundation for Statistical Computing, Vienna, Austria. ISBN 3-900051-07-0, URL http://www.R-project.org. This website contains a downloadable statistical software tool for computing and graphing statistical calculations.

"StatCrunch, Statistical Software for Data Analysis on the Web." http://www. statcrunch.com. This award-winning data analysis package runs entirely in a web browser. StatCrunch includes most features that would be used in an introductory statistics course.

